TECHNOLOGY: A SECOND LEVEL COURSE

T204 BLOCK 1

DESIGN

PRINCIPLES AND PRACTICE

BLOCK 1

AN INTRODUCTION TO DESIGN

PREPARED FOR THE COURSE TEAM BY
CATHERINE COOKE

T204 DESIGN: PRINCIPLES AND PRACTICE

CONTENTS OF THE COURSE

The Open University, Walton Hall, Milton Keynes, MK7 6AA.

First published 1995. Reprinted 1997 with corrections.

Edited, designed and typeset by the Open University.

Printed in the United Kingdom by Henry Ling Limited, The Dorset Press, Dorchester, Dorset DT1 1HD.

This text forms part of an Open University Second Level Course. If you would like a copy of *Studying with the Open University*, please write to the Central Enquiry Service, PO Box 200, The Open University, Walton Hall, Milton Keynes, MK7 6YZ, United Kingdom. If you have not already enrolled on the Course and would like to buy this or other Open University material, please write to Open University Worldwide, The Berrill Building, Walton Hall, Milton Keynes, MK7 6AA (Telephone 01908 858785; Facsimilie 01908 858787; E-mail address ouwenq@open.ac.uk)

ISBN 0 7492 8455 2

Edition 2.1

CONTENTS

STUDY GUIDE

AIMS

This Block aims to take design apart, to reveal and analyse the range of factors that influence the forms and characteristics of designed objects, and to build up your understanding of how these very different factors are brought together in any act or process of designing.

INTRODUCTION

The Block has three parts.

Part 1 is called '*What's it all about?*' It places the design of objects within the larger context of other decisions we make about our affairs and our material environment. It looks at the way objects emerge and evolve within a culture, and at the kinds of situation which lead to the creation of a new object or type of objects. Homing in onto the designer, it then discusses the kinds of knowledge he or she needs, and the historically changing position of a designer as a link and mediator between the makers and users of objects in our society. It looks at the different means of communication designers use to transmit their ideas and proposals to these other groups of people. Finally it examines various issues that arise when we try to judge whether or not a particular design is 'good'.

Part 2 is called '*Let's sit down for a while*', and as you might imagine, it concerns objects for sitting on, more specifically, chairs. Taking a wide range of examples from various countries, most of them designed and produced during this century, it takes the main issues of Part 1 into greater depth by asking 'why are all these chairs so different?' The chairs are selected and grouped into seven sections which between them explore how the main factors influencing design were present in different proportions in each case, reflecting the different human priorities and technological possibilities of the different situations that caused them to be designed and made. On one hand the factors identified here include various kinds of practical, psychological and symbolic requirements of human beings. On the other, they include the constraints and opportunities created by the particular raw materials available at given times, and by advances and innovations in production processes in industry.

Part 3 is called '*Recap and conclusions*'. After the detailed examples of Part 2, it returns in greater depth to questions of how the different participants in the overall design process relate to each other, and of the overall logic of economic and information exchange, of different social roles, that links their various different priorities and contributions together. It takes you back into the material of Part 2 to extract certain kinds of information as the basis for an 'encyclopedia of possibilities' from which the TMA Activity, continuing this process, will lead you to design a new chair for yourself. This Part ends by 'Bringing all the factors together' into a simple network of links between all the main topics and participants. In one direction, this network diagram summarises the whole Block. In the other, it sets up the framework for the highly structured Chair Design Activity which forms the main part of the Block 1 TMA, and enables you to get first-hand experience of the problems of linking together the many diverse fields which impinge upon design.

A checklist of learning objectives is included at the end of the Block. Use these actively in your study and revision of the Block, as they highlight many important ideas and concepts which you may not have noted in your first reading of the main discussions and narrative.

WHAT YOU HAVE TO DO

This Block comprises:

- the *Drawing Work Book*. Whatever your previous experience in drawing this will develop your skill in free-hand design sketches. If you have no previous experience at all, you can start this book immediately and work through it in parallel with your study of this text.
- this main text, which includes self-assessment questions (SAQs).
- a video-cassette divided into two parts, together with a study guide. The first short part relates directly to this Block; the second, much longer part, is a tutorial on more formal drawing techniques.
- a TMA which consists of some brief written questions and the Chair Design Activity, for which there are detailed instructions and a special pack of equipment.

First of all you should read the main Block text called *Introduction to Design*, doing the SAQs as you go. This will lead you very smoothly into the Chair Design Activity which forms the larger part of the TMA.

Somewhere towards the latter part of your work on that Activity – whenever you feel like a change of work for a while – watch the first bit of Video 1, called 'People and Mechanisms'. This lasts only about a quarter of an hour, and gives you a concept through which to look at objects, whether to analyse or to design them, that is of general use throughout the course. It might add a further dimension to your thinking on the Activity, but if you choose, you can leave this short video sequence until you have finished the Activity completely. You will need to have studied it before attempting the written part of the TMA, however, as one of the questions there requires that you apply the concept in it.

The second and longer part of the video contains your first tutorial on formal drawing techniques. This demonstrates how three-dimensional objects are rendered accurately onto the two-dimensional surface of a drawing by various kinds of geometrical procedure called 'projections'.

An understanding of these and some practice in them will be necessary for reading drawings and making your own drawings later in the course, but are not needed for the Activity here in Block 1. There is nothing to stop you studying this video in parallel with the work on this Block, if you wish, but you might find it more inhibiting than helpful for the simple sketching required in the Chair Design Activity. So the recommended approach is to leave this second bit of video till after you have finished the TMA, then study it before moving on to Block 2.

Note that the study guide for the video-recording and the instructions for doing the Chair Design Activity are in the white book called *Supplementary Material for Block 1.*

STUDY CHART

Part	Main Text	Activities	Video	Audio
1	What's it all about?	Drawing Work Book		
2	Let's sit down for a while			
3	Recap and conclusions	'Chairs and motivations' Exercise		
		Chair Design Activity TMA 01 Part 2		
		TMA 01 Part 1	Video 1, Section 1: 'People and mechanisms'	
		Home Kit drawing equipment and materials	Video 1, Section 2: 'An introduction to drawing'	

PART 1

WHAT'S IT ALL ABOUT?

1 WHAT DO WE MEAN BY 'DESIGNING'?

Before we start, it is worth pausing to look at the real meaning of the verb 'to design' in order to clear away any limited preconceptions we have. A standard Oxford dictionary will tell us that 'to design' is 'to conceive a mental plan' for something. This immediately gives us a very important observation: design is a mental process. One thing that distinguishes us from animals is our capacity to conduct quite conscious and complex mental analysis. We can abstract what we learn from the outer environment through our senses, and we can process the information purposively and predictively. We can make plans or designs and carry them out over the long term: we can 'think ahead'.

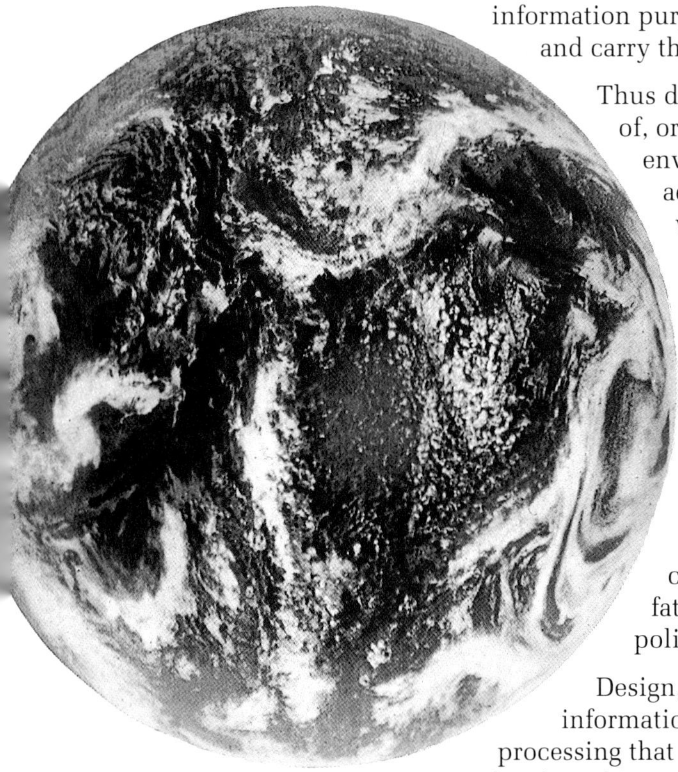

FIGURE 1
'PLANET EARTH', SEEN FROM SPACE

Thus design signifies intelligent and imaginative manipulation of, or intervention in, our material and non-material environments. In today's world where so much control and action in society is conducted and exerted through forces we cannot see, like information or energy, it is important to understand how the same idea of 'design' operates within the non-material world as well as amongst material objects. If you know anything about the ideas of Systems Theory, whether from the Technology Foundation Course, or perhaps from more advanced courses from the Systems Discipline, you will be aware how a system that is entirely non-material is also something that is designed: a management system, for example, or an information system. We can design legal systems. Self-conscious formulation of a political constitution like that conducted by America's founding fathers is likewise precisely a process of designing a political system.

Design, like all human planning, consists of assembling information about a present state of the world around us, processing that information according to the models we have in our heads and in our culture about how that world works, and making a plan of action for some kind of intervention in the situation. The intervention may mean either modifying what is there at present, or introducing something quite new into the situation. Either way the intervention may take the form of a purely organisational change. Or it may involve the introduction of some new object or set of objects into the physical environment. Or it may be a combination of the two: some reorganisation and a new object.

Let's look at this a bit more closely.

2 IS DESIGNING AN 'OBJECT' ALWAYS THE RIGHT ANSWER?

In this course, our specific concern is how the process of design helps us make useful interventions and changes in the *material world of objects*. You may ask why, then, are we talking about these vaguer fields at all, instead of sticking to that specific topic of material objects? The reason is simple. It is what Kipling said: 'What do they know of England who only England know?' Or Lenin, if you prefer the opposite end of the political spectrum: 'In order really to know an object it is necessary to study all its internal and external connectivities'. Both these are historical expressions of what we would now call the systems view: that in order to understand anything significant about a situation at the level of principle, or to analyse it, it is necessary to step outside it.

If we just want to learn to make different versions of say, a table, we can achieve quite a lot by operating as the traditional craftsman did, knowing only about tables and some other kinds of furniture, and never significantly varying the structure or form of what was produced. But if we want to break out of a routine of what is essentially repetitious reproduction of the same object, we need to understand some larger principles. Only by such principles can we open up new avenues of exploration that give us some certainty that we shall not waste time and material fruitlessly, and that we can constantly offer society better solutions as its requirements change. Thus we need to understand what it is we are doing.

At one level we need to see our example of 'tablemaking' in the context of other areas of craft. Perhaps there are things to be learned about tables from examining chairs, say? Then we need to examine whether a 'table' in the conventional sense and form is really the best answer to the problem which our user or customer has. This imaginative rethinking is one of the designer's main responsibilities. It is also one of the reasons why the person who buys a designer's services is more than just a 'customer', and is usually called a 'client'. A *client* is someone who pays for a service, that is for a two-way process tailor-made to that person's needs, whereas a *customer* buys some standard and ready-made object. When someone with a need approaches a designer, the object (or whatever the 'answer' may prove to be), does not yet exist. It will emerge from the dialogue between them. And that dialogue may involve asking some questions which are very fundamental.

To serve the client's needs to the client's greatest advantage, most professionals see it as their obligation not just to answer the question as posed by the client without further probing; not just uncritically to follow conventional preconceptions about how best to solve a problem. The professional's role is to probe a bit deeper, and to help the client question whether those first assumptions were actually the right answer to the problem, before any money is committed to a particular 'solution'.

This is what professionalism is largely about. Compare the situation of the doctor, for example. If a doctor assumes the patient's own diagnosis of his or her illness to be right, and simply provides the pills requested without further analysis to check the diagnosis, we would consider that doctor highly irresponsible. Of course the patient, like the client, may indeed have made the right diagnosis of his or her own problem, but even in a seemingly simple case, it is the obligation of the professional to examine the symptoms objectively and use his or her professional knowledge to advise appropriately. Pink pills may indeed prove to be the best answer in the circumstances, but the doctor must verify that, not just accept it as 'obvious'. Likewise the designer must verify that a table really is the best answer to the underlying problem with which, as the doctors say, the client 'presents'. Perhaps what the client needs is a part-time secretary to clear his paperwork faster, not a table for the papers to pile up on? Perhaps what the patient needs is a holiday as respite from caring for the aged relative, not pills with which to suppress the stress?

It is worth noting the dilemma this can create for a designer – which is shared by the busy doctor. Whether the designer is a single individual faced with a single client, or a member of a team in a design office whose client is a vast international conglomerate, there can be a moral dilemma here. Should the designer or designers take the line of least resistance and minimum effort: give the clients what they ask for, take the fee and get on with the next job? Or should the professional expend considerable mental effort and time, for which most clients are unwilling to pay, in researching the problem afresh and thinking it through thoroughly – only perhaps to come up with a confirmation that the answer the client first thought of was right? In an industrial design situation, say, should the designer provide a quick fix, a superficial restyling of a product which is becoming slow to sell, for example. Or should the designer re-examine the client firm's whole production and employment strategy in order to secure their commercial viability for another five years, when the management itself does not want to bother with such a long-term view? This is one of many areas in which the designer's own value system may come to play a considerable role in the definition of the problem, just as it may in shaping the proposed solution.

Thus at one level, in our example, the designer must help the producer to look beyond the habitual assumption that the market needs tables of a conventional kind, and to see that the 'table' might be enriched or modified into something more useful to a new situation by introducing some new ideas into it. The result might be a table of an entirely new kind, but it is still a table, and of course still an object.

At another level, however, analysis may show that the client's problem is much better solved by some means that does not involve producing an object at all. Reorganisation of some *non-material* aspect of the situation might solve the problem much better. Perhaps a different timetable, a different management routine, a different approach to calculating a budget, for example, are better ways to solve the problem with which the client first approached the designer. Instead of designing railway carriages which jam even more passengers into the space, perhaps a change in the fare structure would distribute the passenger load more evenly during the week and the day?

So we have two kinds of solution which a designer may advocate. There is plainly also a third, mixed type. The solution may be a combination of the material intervention, i.e. the new object, with non-material intervention, such as the change in information flow or pricing.

The other broad category of option, as you may already have spotted, is to design or redesign something, but in a different part of the overall operation from where the client thought he wanted changes. Here is a classic example: the government department poses the problem of how to design a power station that reduces carbon dioxide emissions by producing more power from a given amount of fuel. That is probably a perfectly feasible engineering task, but is it the shortest or cheapest route to cutting carbon dioxide emissions from power stations? A better answer to the actual problem might be to devise easily-installed insulation systems for people's houses so that far less fuel need be consumed in the first place.

What matters is to understand that there is a complete continuum here, in which 'new objects' are only one of a number of possible strategies, or elements of a strategy, for solving the problems that people bring to designers. This view of objects as part of a larger system, of which some elements are material and others are non-material, is not only important to the understanding of options in a particular design task; it is essential when we come to see, in later Blocks, what design does and how it is used in complex situations in industry. It is from this wider environment that the 'brief' for any product or object emerges, and it is that environment which sets the boundaries and constraints upon the problem when it does actually come to the attention of the person we conventionally call the 'designer'.

3 INCREMENTAL OR RADICAL CHANGE?

Whichever kind of solution we choose, material or non-material, or whatever combination of them, there are basically two possible patterns of change or innovation which can take place. There can be big, dramatic, leaps, or small, nearly imperceptible, steps.

The former is called 'radical', from the Latin *radix*, meaning root. By going back to the roots of the problem and thinking it through afresh, a solution is devised which is dramatically different from the previous one.

The latter pattern of change is called 'incremental'. 'Increment' also comes from a Latin word, from the verb *crescere*, which means 'to grow'. Like natural growth, incremental change involves many small steps each of which seems trivial at the time, but whose cumulative effect over a longer period is very significant.

Progress and advance in the world of artefacts depends very largely on people making small modifications to established 'types' of object in this step-by-step way. If you look around your house and think about how certain things 'used to be', examples will spring to mind. Take the kettle for example: it started as being no different from any domestic cooking pot, with a near-spherical bottom for maximum heating surface in the flames of a fire, and as a result could only hang, not stand on a flat surface. On a larger scale, the domestic 'copper' for water heating was an example of the same principle. For the pot to stand on any flat surface, legs were required, as the first two pictures here show. Then when cooking fires were enclosed into 'stoves', the bottoms of kettles were made flat, for maximum heat absorption from the top of the stove. Kettles for use on Aga-type stoves are still like this. When the electric immersion-heating element was developed to heat the water from inside the kettle, rather than an outside source, the kettle form remained much the same. Only in the last few years – and you will have seen it yourself – did someone tackle the problem of the inefficiency of this form for heating only a cupful of water. So kettles started to be made tall and thin, with a heating element in the bottom that is fully covered by only a mug-full of water. This shape is slower for heating a couple of litres of water, but since domestic kettles these days are predominantly used for making odd cups of tea and coffee, this shape has become popular. In this example, the preceding form has not been totally superseded, as it remains more appropriate for certain uses, but any look in the household appliances department of a shop will show you that the new type is now about half of the market. Most of the traditional and well established artefacts in our culture developed through this kind of incremental change. The farm wagon is one example much studied by design historians, and there are innumerable others.

Sometimes the result of quite a small incremental change is not just to produce another example of the same 'type' of object, but to produce something new – on the evolutionary analogy, another 'species', sufficiently different to warrant a new name. Going back to the table, for example, we can easily see how it evolved into the domestic or restaurant 'trolley'. That is just a single evolutionary step: put wheels on the bottom of the legs, replace the horizontal stiffening bars by a flat 'tray', and a table became a 'trolley'. The following example from transport is more complex, because it shows both incremental change and radical change at different stages. It also shows how much inertia of habit often has to be overcome to achieve changes which in retrospect look 'obvious'; how our conventional 'models' of a problem can hold back their real solution.

(A)

(B)

(C)

(D)

(E)

FIGURE 2

(A) ROUND-BOTTOMED COOKING POT WITH FEET HANGING OVER AN OPEN FIRE: DETAIL FROM A 16TH CENTURY VIEW OF A KITCHEN BY PETRUS VAN DER BORCHE

(B) ROUND-BOTTOMED COOKING POT WITH FEET STANDS OVER FLAMES ON AN ENCLOSED KITCHEN RANGE: A MID-19TH CENTURY KITCHEN. DETAIL OF ENGRAVING 'LA BONNE MÈRE', AFTER A PAINTING BY MICHEL–PHILIBERT GENOD

(C) FLAT-BOTTOMED KETTLE FOR STANDING ON HOT FLAT TOP OF AN ENCLOSED KITCHEN STOVE OR RANGE: LATE 19TH CENTURY

(D) WIDE-BOTTOMED ELECTRIC KETTLE WITH INTERNAL IMMERSION-HEATING ELEMENT: RUSSELL HOBBS LTD., K.3 MODEL, LATE 1960S

(E) NARROW-BOTTOMED 'JUG' TYPE ELECTRIC KETTLE: HADEN AUTOJUG, MID 1980S

FIGURE 3

(A) AN EARLY DEVELOPMENT OF THE DIRECTION-INDICATING MECHANISM TO REPLACE THE HUMAN ARM STUCK OUT OF THE SIDE OF THE MOTORCAR: BRITISH, c. 1913. THIS LITERAL 'MECHANICAL HAND' WAS A STANDARD ACCESSORY THAT COULD BE PURCHASED AND FITTED TO ANY CAR. MADE OF THE EARLY PLASTIC BAKELITE, IT WAS FITTED WITH A LIGHT BULB FOR NIGHT USE. THE CONTROL UNIT WAS CLAMPED TO THE STEERING WHEEL, WHENCE A CABLE RAN TO THE HAND CLAMPED OUTSIDE THE CAR'S OFFSIDE. THIS NORMALLY HUNG LIMP IN THE VERTICAL POSITION, BUT COULD BE FLIPPED UP OR TWISTED IN IMITATION OF A HAND SIGNALLING RIGHT AND LEFT TURNS, BY OPERATION OF THE CONTROL KNOB

(B) ABSTRACT REPLACEMENT FOR THE HUMAN HAND: THESE 'FLIPPER' INDICATORS WERE FITTED TO BOTH SIDES OF THE VEHICLE. A TYPICAL EXAMPLE OF THE 1930S AND '40S ON A ROVER 60 OF 1948

(C) EARLY VERSION OF THE FLASHING DIRECTION-INDICATING LIGHTS AT THE CORNERS OF THE VEHICLE: DAIMLER 'DART' CONVERTIBLE, 1959

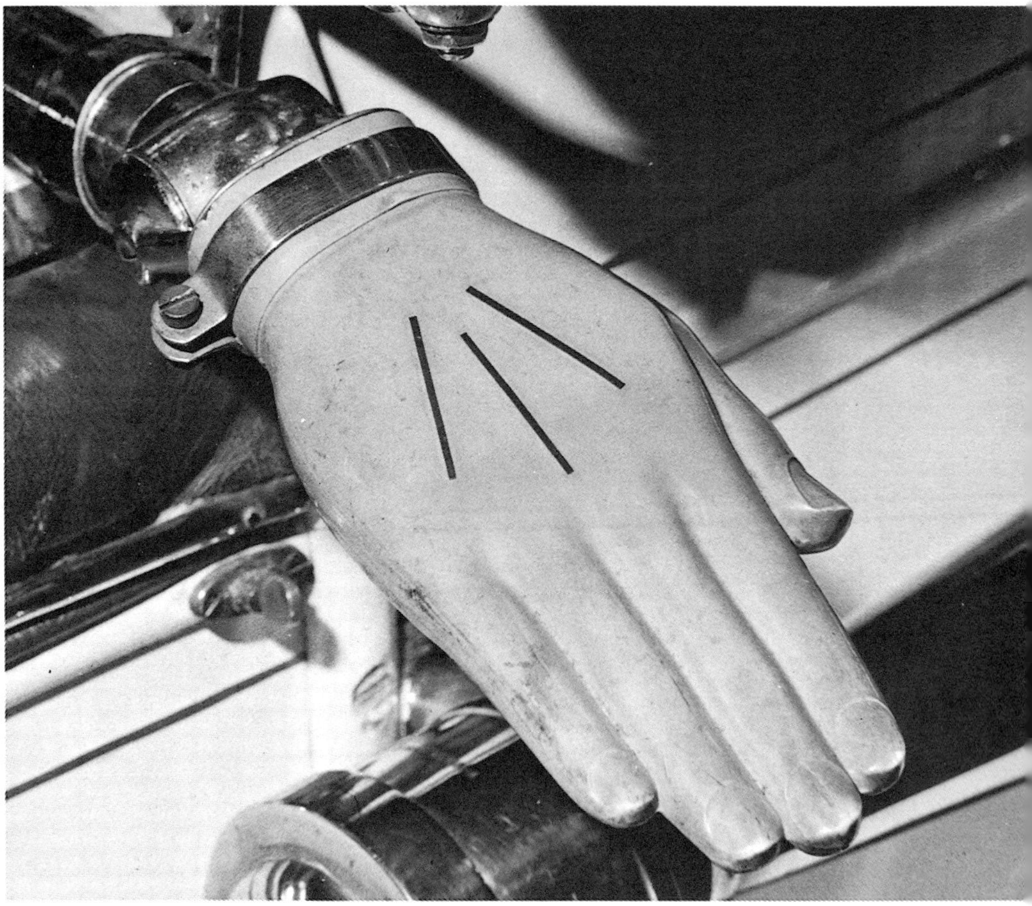

(A)

(B)

(C)

When motor cars were first developed, in the early years of this century, there were of course very few of them on any given bit of road. If one driver needed to indicate to another that he or she was about to turn, they would stick the right arm straight out (in this country, driving on the left), to indicate a right turn, and make a wheeling movement with the right arm if indicating they would turn left. Then cars started to be enclosed with solid roofs, and between the two World Wars, driving became a form of mass transport rather than just a rich person's recreation. Getting the arm out of the very small windows of those days was awkward, and if raining, very disagreeable. So the human arm became replaced by a little mechanical arm with a light in it, which flapped out like a penguin's flipper from the central upright on the left or right side of the car, as appropriate. For anyone with narrow gateposts, they could be a nightmare: if the driver did not remember to bring them in again quickly, they regularly got broken off.

Through much of the 1950s, car indicators were still like this. If you are under about 35 you may never have seen these extraordinary objects. And you will justifiably marvel that it took so long for someone in the motor industry to realise that 'indicating' did not have to be an exact mechanical simulation of the human arm; and moreover that in denser traffic, the message was better communicated from all four corners of the car, not just from a position in the middle of its side that was obscured from most viewing angles by the car body itself.

That example is a nice combination of the two different kinds of change. First there was a very small evolutionary or *incremental* step: a human arm was replaced by a little mechanical arm in an almost identical position. (Very occasionally, as in the illustration here, this was literally a 'hand', but usually it was an illuminated flipper.) Then there was a *radical* step: someone looked at the underlying problem afresh; they saw it was a problem of 'communicating directional intentions' not of 'making a substitute arm'. The result was quite a different kind of object in an entirely different location.

The resulting flashing corner lights in turn then went through their own process of evolution. Initially, like the headlamps of that date, they were separate objects in their own right attached to the outside surface of the car body. By a series of incremental changes they became increasingly more streamlined in form, better weatherproofed in their detailing, and less vulnerable to damage as they moved towards the form they have today, where they are no more than one element of the jigsaw which makes up the smooth outer form of the car, totally absorbed within its surface, and taking their place amidst what are now several forms of 'message-emitting lights'. This question of the two main kinds of change is very important to understanding how a designer may work and what the results of his or her work may be. It is explored in greater detail in Block 3.

4 WHERE DOES THE IMPULSE TO DESIGN SOMETHING COME FROM?

As we have already observed, the pressures and demands which shape an object in any situation are usually very complex. This is because objects serve very many different kinds of needs in society and in the lives of individuals, because groups of people have differing priorities. These differences are not just the result of practical priorities, in a functional sense, in that people have quite objectively differing practical needs. It is also a matter of the mass of different value systems in our society which cause people to give quite different weightings or priorities to different factors or aspects of our human existence.

In societies which operate more or less at subsistence level in terms of goods and services, any practical solution or workable product is welcomed, and its very possession prized. The affluence of choice to which the West is accustomed, however, produces a highly differentiated and selective market in which no one design solution for any type of product is likely to satisfy the whole population. Even groups which have identical practical requirements will be looking for subtly different symbolic or aesthetic characteristics in the object concerned. This is particularly true of consumer goods, but it is more true than one might initially suppose of certain industrial goods, which also have certain psychological roles to play amongst their users. Research has shown that even in fields like scientific instruments and medical equipment the aesthetics of the objects play a significant part in influencing their purchasers' choices.

Such extreme diversity is a relatively new phenomenon. Earlier periods of history when there was typically a more coherent and uniform belief system within a given society – thanks not least to the role played by religions – priorities were more uniform in relation to objects. Take clothes as an example. At any given time from the medieval period in England and indeed right up to the late fifties of the present century, there was a more or less uniform agreement as to what constituted appropriate clothing for everyday life. 'High fashion' was an indulgence of the rich few, and was one of their modes of power display to the less fortunate. With the growth of a multi-cultural, highly differentiated society since the early sixties, there is no such general conformity or agreement running right across society. The everyday clothes of one cultural group contravene taboos and conventions of another with impunity. A myriad different cultural groups operate throughout society according to entirely different sets of social priorities; they express those priorities and value judgements in their clothes; and their clothes in turn become one of the most powerful media through which they announce those values to the world, and display their membership of a certain social group. Lengths of skirt, degree of explicit body fit or exposure, colour schemes, use of 'precious' materials like gold or of 'practical' materials like denim or tweed – all these and many more are the parameters of statements about the wearer's membership of social groups and personal value systems.

Clothes are a good example of how there is no single 'right' answer to design problems in the abstract sense, only 'right' answers in relation to highly specific sets of social norms and conventions, or in relation to practical exigencies like climate or function.

Now if you have got the message of this discussion, that word 'function' should have jarred a bit with you there. Isn't all that other stuff about social differentiation through clothing also very much to do with a kind of 'function' that clothing performs? Of course it is. It is a very important function: that of conveying messages to each other in an economical but very subtle code, messages which help the process of finding congenial social contacts amidst the vast mobile mass of modern society. Here we see 'function' starting to have a wider definition than the purely practical one which is too often assumed.

Clothes are an example where this concept is relatively well understood by most people. If you actually ask people for honest statements of why they bought certain other particular items in their homes and offices, and they really do answer honestly, it emerges, of course, that very many of those decisions are also made at least partly on grounds of some kind of aesthetic preference. What does that mean? It means that the object gave them some kind of satisfaction through 'what it said', to them or to people in their social circle, rather than just through 'what it did'. The opportunity to make such choices is indeed the result of a certain level of affluence in society: much of the world would be grateful for any kind of electric kettle, let alone the choice between a 'jet-age chrome-plated' model and a 'traditional, farmhouse floral' one. But the fact is, as anthropological information from any culture shows, that this determination to differentiate ourselves into social sub-groups by forms of personal display is absolutely fundamental. The messages will be carried by simple means where the sophisticated are not available, and the forms that carry the messages acquire powerfully symbolic charges to the individuals and groups concerned.

As many a failed product has shown, the balance of a potential user's priorities has to be very accurately judged if a product is to succeed in a market place where there is choice. On the one hand this requires information about what people perceive to be the *practical* shortcomings of products currently available, and how much they would be prepared to pay for a product which overcame those deficiencies. On the other hand it requires information about shifts in the population's *psychological* expectations. Plainly any manufacturer trying to increase a company's sales needs to understand these factors and how they are changing within the target population. Recent decades have seen increasing attention to 'market research' in order to provide this information and help producers to make confident predictions. We shall take this topic further in Block 2.

Returning to the larger situation, it is clear that for any commercial manufacturer, the first and over-riding aim of any new design or innovation has to be something that sells. But there is another side to the equation: there are also needs and requirements on the producer's side that have to be met, pressures from that side to which the strategy must respond. The range of ways and points at which a company can innovate in a given market will always be relatively very wide. In examining alternative possible 'gaps' in the market to fill, or in deciding to go for change at all, there may equally be factors on the producer's side that create the impulse to design and produce a new product of a particular kind or in a particular way.

What kinds of factors might these impulses from the producer's side be? Like the factors on the demand side, they can be either technical or human factors. Maybe an economic source of a key **raw material** has dried up, some other material has suddenly become available very cheaply, or the other firms competing in that market place are beginning to undercut by use of some entirely new **technique**. Maybe production **machinery** is worn out and replacement is the moment for an overall updating that means changing the product. Or in the **social** sphere, suppose another major local employer suddenly closes down and releases a workforce with very specific skills for which our producer sees himself able to provide a livelihood by expanding into a new area of his particular field in a way that uses those particular skills. Or maybe the company's **financial** problems mean that it must get a new product selling in the market by a certain date in order to survive, and the speed required determines all its technical options. Or the well-established product may be overtaken by new **legal** requirements, perhaps of safety or of permissible pollution levels. Or it may simply be that the designer or someone else in the company has had **a new idea** which is so good, in its cost savings or its market appeal, that development costs will be amply repaid by increased profits.

Any of these reasons, and infinite numbers of others within the same categories, may lead to a producer directing his design or innovation policy in a certain particular direction. You can imagine lots more such scenarios for yourself. The point is that this side of the situation will generate specific requirements or constraints which the designer of the new product must take into account just as seriously as the requirements and constraints of the market in which the company hopes to sell it. As the jargon goes, the innovation process is to some extent 'producer led', and to some extent 'market led'. These two factors are sometimes called 'production (or technology) push' and 'market pull'. The development of any particular product often results from a combination of both factors (Block 2 discusses this further).The initial impulse to design a particular kind of new object may come from either side, and the new product will have a function in the economic and social life of producers as well as consumers. The task of the designer is to devise something which solves the requirements of both sides simultaneously.

5 WHAT KINDS OF KNOWLEDGE DOES THE DESIGNER NEED?

No amount of market research, financial or production information actually tells a designer how to arrive at an object which answers that combination of requirements. What such information does tell a designer is something about how the object is to be produced and the properties which the finished product must possess. The actual process of arriving at something which is simultaneously a good plan of action for the company and a good product for the customer is of course the process of 'designing'. How this process will go, in terms of the steps the designer will take, the line of thinking he or she will pursue and so on, is partly a matter for the person concerned. It is also partly a matter of established kinds of logic and procedure which have been shown through past experience to be efficient working methods for resolving this kind of problem. We shall look at this question in more detail later in the course.

What I want to point out now is that there are several quite distinct things here.

First, we can divide the whole situation into two: the 'problem', and 'what the designer brings to the task of solving it'.

The problem will be defined by a lot of 'information': all that stuff which market research, finance and production people can provide.

'What the designer brings to the task of solving it' divides in turn into two again. Firstly the designer has 'knowledge', derived from learning and experience. Secondly he or she has a 'technique' or 'skill' for the highly complex process of resolving information and knowledge into a good set of proposals for what should be made or done. In both these areas, *knowledge* and *technique*, there will inevitably be much in common between different designers, but also much which is individual. The individuality derives from a particular professional background up to that moment and from a personal value system. As with any human being, these two will be closely interrelated and usually have a mutually reinforcing effect, as value systems incline the individual to seek certain kinds of experience, and experiences shape the value systems. The extent to which a designer allows these personal values to have an obvious impact on how design is conducted and on the characteristics of resulting designs is a matter for the individual in each case.

Long before environmental issues became so topical, for example, there were designers who would only design in accordance with energy- and resource-saving priorities, and whose whole careers and reputations are based on pursuit of that ideal. Two famous examples were the American architect-engineer the late Richard Buckminster Fuller, and the still-active American industrial designer Victor Papanek, both of them vehement critics of mainstream industrial culture in the developed world. Now that energy and resource issues are a widespread concern, there are many designers who will adopt 'green' priorities if those are the client's preferences, and be quite prepared to design to a ruthlessly commercial set of priorities if that is another client's requirement. Probably most professional design consultancies today would come somewhere in that middle category.

At the other extreme of this particular set of values, there are designers who have no interest at all in this kind of social conscience or environmental awareness, who are addressing the developed world's marketplace in its own terms and even treating its absurdities and foibles as part of the raw material to exploit financially, or to play games with

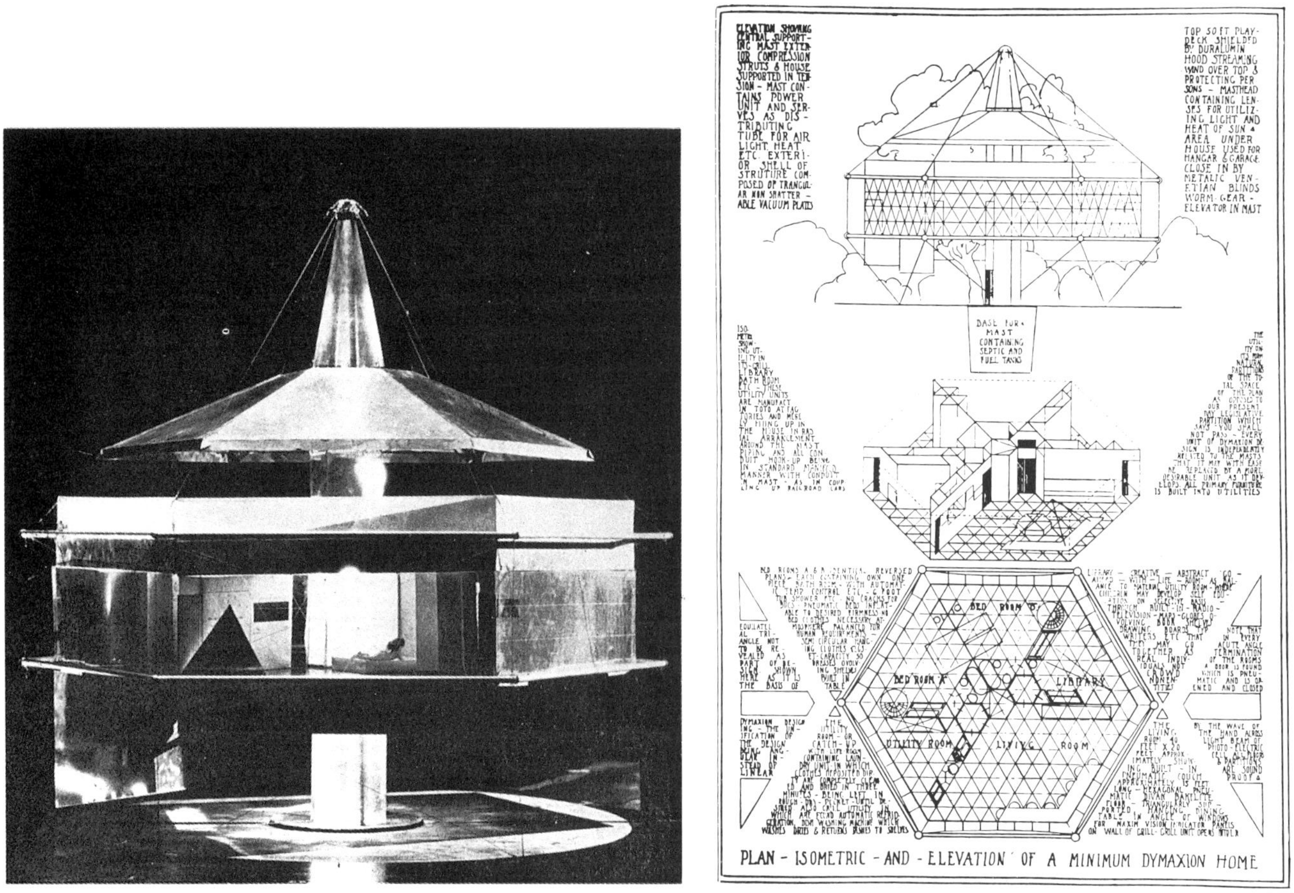

FIGURE 4
RICHARD BUCKMINSTER FULLER, 'DYMAXION' HOUSE, 1927–9: PHOTOGRAPH OF THE MODEL AND EXPLANATORY DRAWING

FIGURE 5
VICTOR PAPANEK AND STUDENTS, STOCKHOLM, MUSCLE-POWERED MACHINE FOR TRANSPORTATION OF GOODS, STRETCHERS, ETC., IN UNDER-DEVELOPED AREAS, MID 1960S: PHOTOGRAPH AND DRAWINGS IN OPERATION

FIGURE 6
ETTORE SOTTSASS, 'CARLTON' ROOM DIVIDER IN BRIGHT COLOURED PLASTIC LAMINATES, FOR MEMPHIS SpA, ITALY, 1981

aesthetically. One could name many here, but for many years the most influential and provocative has probably been the Italian designer Ettore Sottsass.

There are thousands and thousands of professional designers all over the world whose personal value systems lie somewhere between the extremes. As when buying any other professional service, a client tends to look for a designer sympathetic to his or her own priorities as well as one with expertise relevant to the task. But in hard times particularly, the matter of striking a balance between financial survival and 'fighting for what we believe in' is just one more of the difficult issues a designer has to resolve afresh on every job.

From our discussion so far, you can begin to see why this 'process of designing' is so complex. The factors which the designer is having to reconcile and juggle are of totally different kinds. They do not just come from different areas of knowledge; they are qualitatively of quite different orders. At one end of the scale of certainties there are the minutely exact facts of engineering science and materials behaviour. In the middle are relatively uncertain, but hopefully more or less predictable matters, such as production costs and component availability. At the other end are largely subjective and intuitive judgements on how unknown and fickle persons will respond to certain design features, shapes and colours in a given culture in, say, two years time.

Producing a coherent and responsible strategy out of all that, when many people's livelihoods and pensions depend on your answer being a success in the marketplace, is not a task susceptible to complete analysis or to simple specifications of 'how it is done'. This is not least because in every case, the balance of priorities is different. Much depends on what generated the initial impulse in the first place. Within specific areas of design where priorities are relatively constant, the logic of the process can be more or less formalised and the 'design process' reduced to a 'method' or 'routine'. Much of engineering industry, for example, works to some pretty standard sequences whereby first this question is resolved, then that and then another. But the balance between 'science' and 'art' is very variable in design as a whole.

Some jobs that are termed 'design' in fact involve little more than application of standard engineering data tables and standard mathematical formulae to fix the specific component sizes required in a variant of a standard object. Thus one may find the designer in a gearbox manufacturer who produces drawings for a gearbox with certain specified ratios as required by a certain client. This designer does not devise any new layout, apply any new engineering principles, or innovate in any way at all. Many designers in industry are of this kind: the routine is established and 'design' involves producing variants on the theme. This is naturally the kind of design where computers are most rapidly replacing people – and often they don't produce drawings at all, they just pass coded instructions direct to the production equipment to control the precise shape and dimensions of what it produces.

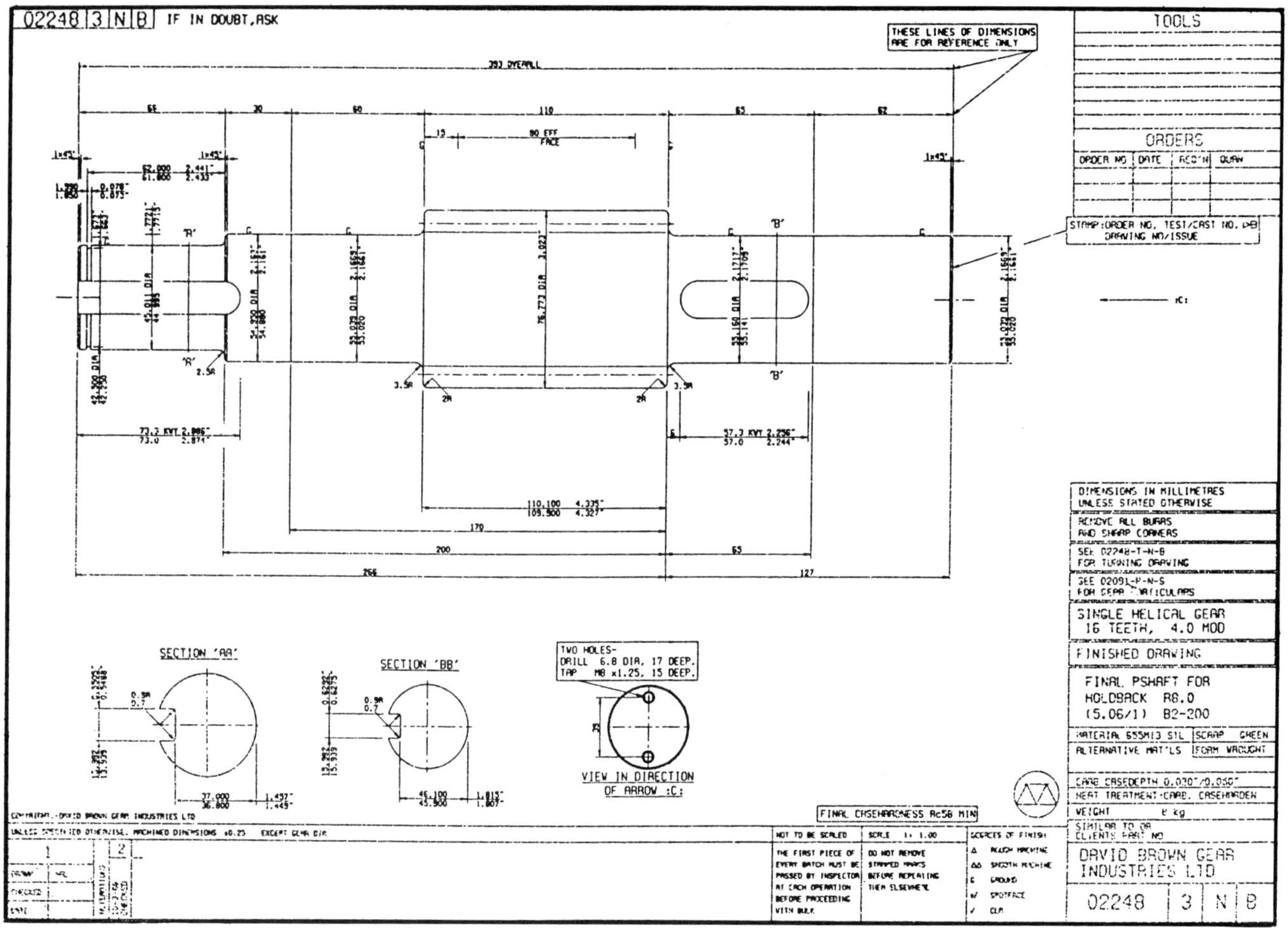

FIGURE 7
DAVID BROWN GEAR INDUSTRIES LIMITED, DESIGN FOR A GEAR SHAFT, COMPUTER DRAWN DIRECTLY FROM THE APPLICATION OF ENGINEERING DATA TO STANDARD FORMULAE WITHOUT INTRODUCTION OF ANY NEW PRINCIPLES AND WITHOUT ANY PRECEDING DRAWING BEING EXECUTED BY A HUMAN BEING

The kind of design that a computer cannot do (as yet) is the kind that invents a wholly new type of gearbox, say, operating on a wholly new principle. The sort of wide-ranging thinking and experience of the world of possibilities that characteristically lies behind such 'break-throughs' or 'inventions' is still the unique province of human beings. The processes of thought that produce such new ideas are not without observable pattern, and quite a lot of what a designer does in training is to practise those patterns of working which are known to help produce good ideas, through a series of design tasks that build up his or her personal experience. We shall look at some examples of these patterns of working in Block 3.

By looking at some case histories of how designers did things we shall try to show, later in the course, how this activity of designing proceeds. Basically it is a process that combines long periods of tedious plodding and rejection of options that do not solve the problem, with brief exhilarating periods of positive progress. But that as I have said is for later. That is the designer's skill in bringing various kinds of knowledge to bear on the problem. Here I want to sum up what those various kinds of knowledge are.

We have already seen that the designer needs to know not just how to manipulate the material environment in which we operate, i.e. how to shape given materials for use as components and objects. He or she also needs to know both 'how the world is' physically, socially and economically, and 'what is important' to the people in it, politically, psychologically or symbolically.

So we have three areas of knowledge here.

First, we have *science*: organised bodies of knowledge which try to describe 'how the world is'. Everyone would recognise that physics and mechanics are relevant sciences in any engineering field. So too is the whole science of the behaviour and properties of materials. Block 5 will touch on these sciences when discussing detailed issues of engineering design and selection of materials appropriate to a given task. Geometry is another science in this sense: our organised body of knowledge about what is possible in three-dimensional space, what the useful properties of certain arrangements or configurations of elements in space may be. Block 4 will look at the science of geometry as it relates to basic design problems. In some respects, statistics and economics are sciences, too, and enter into fields like market research and product specification which are discussed in Block 2. These are elements of the broad category we call sciences, and many factors in these areas can be precisely described, even quantified.

Second, we have *philosophy*. This deals with 'what is important' to people at the most fundamental level. In any particular period, current descriptions of 'how the world is' may influence certain areas of philosophy. Other areas of it operate on principles so deeply embedded in the given culture that no advances in our knowledge of 'how the world is' ever significantly alter them. Many of those deep principles have become incorporated into, and are transmitted through, organised systems of religion. Whatever the origins or lifespans of these principles, it is philosophy, the question of 'what is important to people', which determines the sets of priorities we call politics – that is who controls things in a society for whose benefit – and aesthetics, which concerns what our sensations tell us gives them special satisfaction, or in shorthand, what we consider 'beautiful'.

Third and last, we have *technology*, which is the sum of our skills in handling the material stuff of the physical world around us to serve our own purposes. To some extent it is a marrying of science and philosophy, in that the skills derive from our collective knowledge of 'how the world is', and the purposes are human aspirations that are unavoidably informed by some philosophy.

The science and technology on which members of a given society may draw in any one time and place will be more or less uniform. Thus it is very largely differences in philosophy, and hence in political and aesthetic priorities, which cause us to solve objectively similar problems in very different ways, and that problem-solving includes how we design objects. This is how there come to be many 'different' objects that do what is basically the same job in the same way.

We shall explore all this in more detail later in the Block, when we look at a specific kind of familiar object in relation to these factors.

SAQ 1

What are the main factors which we have identified here as influencing the form of a product:

(a) from the market point of view; (b) from the producer's point of view?

SAQ 2

What are the main kinds of knowledge that the designer must have in order to design?

6 THE DESIGNER'S RELATIONSHIP TO THE PRODUCER AND THE CUSTOMER

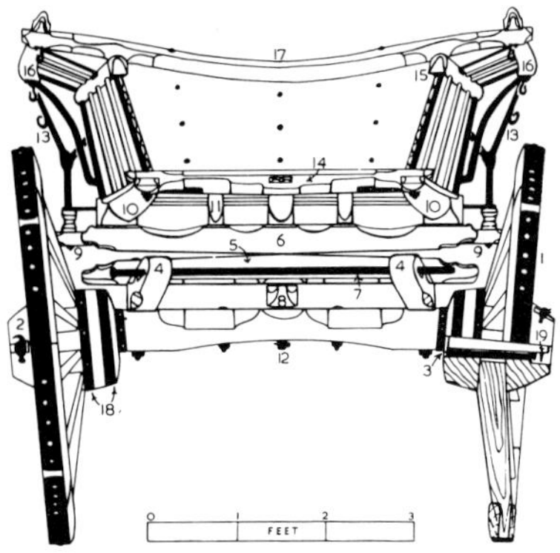

FIGURE 8
ENGLISH FARM WAGON, IN THE FORM TO WHICH IT EVOLVED AND STABILISED, WITH OUTWARD SLOPING WHEELS OVER WHICH THE LOAD-CARRYING AREA EXTENDS, AND OTHER STANDARD FEATURES OF THE VERNACULAR 'TYPE'

In the past, as we have already commented, there was a greater uniformity of value systems within given societies, and of course a much smaller range of technologies available for the production of objects. Handicraft technologies meant that there was generally a direct relationship between maker and purchaser of a product. When populations were small and objects took a long time to make, most things were made more or less to order. Slow changes in technology and relatively constant sets of social values meant that change and innovation in the character and form of the objects which were made in a given society took place slowly. Change generally occurred in small steps, by the process we have seen as evolutionary or 'incremental' change. A given kind of object – say, a table-knife or in the classic example we mentioned, a farm wagon – evolved in form gradually as individual craftsmen slowly improved on a generally satisfactory object. (In those days they were almost universally crafts*men*, not least because most production processes involved some very heavy physical work.)

In such circumstances, it was relatively rare to ask a craftsman to do something radically new. Usually he was asked to produce another example of some well established type of object: another horse's bit, gatelatch, plough, potter's wheel, axe and so on. If it was another teapot, it might be a crafts*woman*, as women often did pottery; if it was a new shepherd's smock, that too probably involved a craftswoman of course. But the characteristic features defining such objects were well established and there was a clear understanding amongst potential customers as to what made a 'good' example of the type, and what was bad. Those who had worked with axes day in, day out for years knew on sight which one was well balanced, would cut easily and would not tire their arms in use. The best form for a teapot spout that poured smoothly and did not drip was likewise well understood. This kind of knowledge was built up through long slow experience, either by the individual concerned, or within a social group which handed its experience on to its children by word of mouth.

When items are examples of a well established 'type' in this sense, and are made by hand to the order of specific customers, communication about what is required, and what will be a good solution, is a direct process. I say 'is', because such situations do still arise occasionally (see if you can think of any examples). The main characteristic of this type of situation is that the object being produced will not differ significantly from a standard type which the craftsman or craftswoman has made many times before. Any differences will be so slight that they can be directly decided between customer and maker without anyone else needing to help them decide by giving 'specialist' advice.

Suppose someone in the middle ages wanted another couple of stools, say, for two new servants to sit on in the kitchen. 'A stool for sitting at the kitchen table' was a well established item. 'A couple of stools, sir?' All the maker needed to agree with the customer was which timber and which particular style of legs was preferred, and the stools would be delivered shortly in return for the cash (or meat, partridges, wine, corn or whatever recompense was agreed).

Of course when the craftsman wanted a stool for his own use, or the craftswoman an overcoat, customer and maker were all one person. That person did not have to utter a word aloud to anyone about the shape or material required. Now as then, communication in such a situation goes direct from the maker's brain through his or her hands to the tools.

Figure 9(A) on page 26 illustrates this. Compare the process of making a routine sponge cake. You know the standard ratio of eggs, flour, fat, sugar off by heart, and it always goes in an eight-inch circular tin. So you do not need any design or manufacturing instructions from a recipe book, nor do you need to verbalise the process in any way to anyone else. More or less literally 'without thinking', you can have a cake ready for consumption in half an hour.

If your neighbour asks you the favour of making a cake for her children, as your cakes are famously good, then maker and customer diverge again to become different people, and words or some other medium for discussion and decision-making must pass between them. Should the cake be made with chocolate or fruit? What shape is preferred? And what size? How much should you add to the cost of ingredients to cover your fuel cost in baking it, as she insists you should? Materials, shape, size, energy costs ... the standard factors in any production transaction are all here. This relationship corresponds to Figure 9(B).

Sticking with these apparently simple examples of the stool and the cake, let us see what happens as customer requirements get a bit more sophisticated.

Suppose that some people with a very small modern kitchen want a set of four breakfast stools, two for parents and two for kids, which will also fold flat either to clear the kitchen space or to enable them to be used in the family caravan at the weekends. Hm! Not such a standard problem that for the friendly carpenter they have found. The carpenter has never made anything to fold flat before, and all he can do is improvise some kind of modification to the standard kitchen stools he usually makes, hoping they will not collapse. It would be very expensive to get them wrong and have to repeat the job. Even worse if the stools did collapse in use and people were hurt. No room for trial and error therefore, and there is no exact precedent to follow for the principle. Worse still, the particular spaces into which they must fit in the caravan impose some tight dimensional constraints. The carpenter has the sense to know that, technical wizard though he is with a bit of mahogany, he is quite out of his depth in devising a strange new kind of animal for this unique set of requirements. The customer's ideas are equally vague: he only knows what the problem is.

Together they have the sense to know what they need – they need a designer. The customer tells the designer what the family's problem is, how much they would like to pay, and reveals the fact that they are off on holiday in the caravan in six weeks time. This means, apart from anything, that the stools will have to be made by the local carpenter who is desperate for work, since no-one else is motivated or flexible enough to produce something against such a tight deadline.

Once more we have a classic design situation: time constraints, predetermined technology and pre-specified manufacturer, as well as quite complex practical and costing requirements.

Now that the requirements have gone beyond those which the established 'types' of object can perform, the customer is having to pay someone to have ideas, as well as paying someone to buy materials and process them into an object on the customer's behalf. From the designer he or she will be buying 'information' and 'know-how', or knowledge and skill. On payment of the fee for that time and expertise, instructions

FIGURE 9
CHANGING RELATIONSHIPS BETWEEN MAKER, DESIGNER, CLIENT AND USER
(A) ONE PERSON DESIGNS AND MAKES FOR THEIR OWN PERSONAL USE
(B) THE TWO PARTICIPANTS IN THE TRADITIONAL CRAFT PROCESS OF DESIGN AND MAKING
(C) THE THREE PARTICIPANTS IN THE CRAFT PROCESS, WHEN A PRODUCT REQUIRING SIGNIFICANT NEW DESIGN WORK IS REQUIRED
(D) THE THREE PARTICIPANTS IN A TRADITIONAL MASS-PRODUCTION PROCESS
(E) THE FOUR PARTICIPANTS IN A MASS-PRODUCTION PROCESS THAT INCLUDES FEEDBACK OR ORIGINAL PRODUCT PROPOSALS FROM ANOTHER SOURCE, E.G. FROM A MARKET RESEARCHER

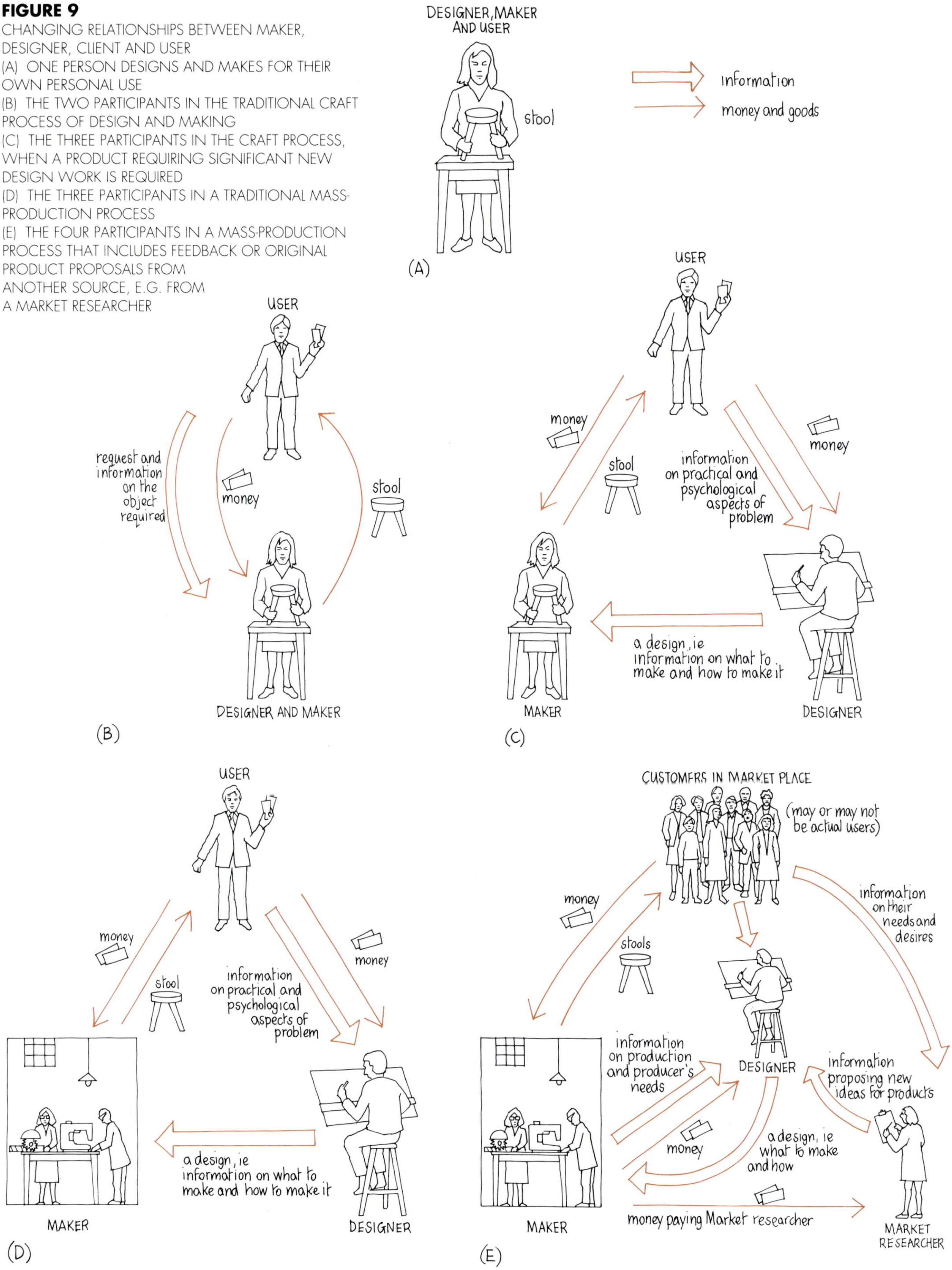

about 'what to do' will pass from the designer to the carpenter. Probably this will be in the form of a drawing, with some further details and explanation accompanying it in words. Figure 9(C) shows this.

Returning to the cakes, we can see a quite different category of problem that also involves introducing someone with specialist knowledge, to mediate between customer and maker as 'designer'.

The practical function of cakes is pretty clear. Where stools are sat on, the practical function of a cake is to give the body energy and nourishment. But 'function', as we have already seen, can be of quite another type. It can be purely psychological or symbolic.

Suppose your neighbour's teenage son is having a birthday party. He was furious a year ago when his mother produced a professionally iced fruit cake from a rather expensive shop, conventionally and tastefully decorated to her special order with his name in blue.

Such a deeply unfashionable item let him down dreadfully in front of his gang, where anything 'conventional' is taboo. Their obsessions seem to change monthly (sometimes faster), according to the latest advertising hype from the pop music industry. Without any birthday cake, he would also feel let down, but nutrition is the last thing the cake needs to provide in this case. As centre piece of the party table, it must do something to confirm his trendiness – but how to achieve that without actually consulting him and spoiling the surprise? As an expert in the current aesthetic, your own daughter advises. This month's hottest number, it seems, is a lead singer whose skin-tight, shiny, purple, leather trousers with provocative green patch send both sexes equally into paroxysms.

Your daughter would never call herself a 'designer', but she threw a recent fan magazine at you where the said pants featured life-size on the centre spread. Dead-smooth shiny purple with an off-centre parallelogram of deep green was clearly the cake that would symbolise the birthday boy's up-to-the-minute fashionability. The act of inserting the knife into this composition produced near hysteria throughout the room. No-one actually ate any cake, but it set the atmosphere of the whole party alight, and established the boy's place as leader of the gang for months thereafter.

Another success for design? No question of it. And not such a trivial example as at first you may think.

The task of creating some psychological attraction between objects and potential customers is a key function of all design for the commercial marketplace. The means by which such vibrations of positive association may be created between customer and object are amongst the central tools of a designer. In themselves, as a body of information on such features as colour, shape, composition, they are value-free, or neutral. In particular combinations at particular times, features of this kind acquire connotations of certain 'values'. It is a basic kind of 'signalling' mechanism that operates here. But another level of values, and hence of moral judgements, comes into play here too.

When the features which cause a certain psychological effect or attraction are applied as an external surface 'styling', to give off some message or create some promise that the object itself does not fulfil, many people who have learned to see 'good design' as something functionally honest in a practical sense will condemn it. According to this point of view, there is something immoral in, for example, streamlining the bodywork of an ordinary family saloon car to make it look like an excitingly futuristic high-speed machine.

It is even worse, in this view, when the objective is to encourage purchases of items for which there is no practical requirement, which are bought simply in order to fulfil artificially generated desires for certain kinds of 'cult status' in social groups. To some, the involvement of any designer here is seen as cynical, even 'wrong'; as a morally offensive squandering of natural and human resources in order to pander to instincts for social competition. According to some value systems, that judgement would be entirely justified. There is another set of values however, which sees the creation of jobs and the satisfaction of human psychological needs as entirely legitimate functions of industry and a legitimate use of these resources in its own right. According to this view, the designer is as honestly employed when 'making people happy' as when fulfilling the practical functions which ensure their biological survival.

These two positions are extremes, on a continuum that involves almost every product or object which is created. As we have agreed, nothing in a sophisticated society is value-free. Thus nothing is 'association-free'. Serious and expensive engineering items, for example, are bought as much for the 'efficiency' connotations of their forms and the 'fashionable' connotations of their colours as for their operational functions. In machine tools or industrial vehicles, just as in consumer goods, this field of 'what the thing looks like' and 'what associations does it evoke' can frequently be a factor significantly influencing success or failure in the market place.

The process only goes 'wrong', morally, or from the customer's viewpoint, when the associations which an object evokes are used to obscure or misrepresent those facts of 'what it does' which are actually important to the buyer or user.

Given that this dimension of design is not of itself pernicious, merely, like all others, one dimension which may be manipulated according to the value systems prevailing, it is a field that the designer must master as essential to his or her professional repertoire. When a product is basically sound, attention to this aspect of it can be a designer's most important contribution.

The extent to which this factor is used positively depends considerably on deep cultural conditioning, and is well understood to vary between countries. In Britain, a somewhat Puritan cultural conditioning has led to widespread mistrust of anything appealing to the senses as 'trivial' and ethically suspect. Take the once famous case of portable typewriters, for example. Back in the Swinging Sixties, as now, typewriters were universally rather clumsy objects in an uninspiring grey. Then in 1969, the Italian designer Ettore Sottsass produced one for the Olivetti company which was sleekly contoured and scarlet with an integral handle like a briefcase. They christened it 'Valentine' and presented it through advertising as every girl's highest fashion accessory. In Italy, and thence much of Europe, office equipment has never looked back. In Britain however, the minority which wants its filing cabinets to be pretty or cheerful still has to pay twice as much for such symbols of sinfulness as for those in traditional grey, and probably always will.

FIGURE 10
ETTORE SOTTSASS, 'VALENTINE' PORTABLE TYPEWRITER FOR OLIVETTI, 1969, WITH TYPICAL ADVERTISEMENTS FROM VARIOUS EUROPEAN COUNTRIES PRESENTING IT AS A SEXY FASHION ACCESSORY RATHER THAN AN OFFICE TOOL

GREAT BRITAIN, 1970–71: 'A VALENTINE FOR CHRISTMAS', GIFT PACKAGING BY CROSBY/FLETCHER/FORBES FOR BRITISH OLIVETTI LTD

GERMANY, 1969–70: POSTER FOR 'VALENTINE' BY J WALTER THOMPSON GMBH, FRANKFURT A/M

ITALY, 1969–70: POSTER FOR 'VALENTINE' BY UFF. PUBBLICITA OLIVETTI, MILAN

ITALY, 1970–71: POSTER FOR 'VALENTINE' BY UFF. PUBBLICITA OLIVETTI, MILAN

This raises the important question of what causes such an object to cost more. Does it have some extra mechanical features that raise its manufacturing cost? No. Has hand labour been involved and some artistic or professional skill in applying the special paint? No – we are not talking here of handpainted decoration that converted the filing cabinet into a unique art work. Nor are we talking of special colour selection by an interior designer or colour-matching to an individual specification. There is no identifiable 'value-added' element resulting from intervention of a designer's skill. There is merely a minority taste in the marketplace, therefore a small production run, and a distributor's nightmare of geographically unpredictable demands for each of the 'pretty or cheerful' colours offered. As the cliché therefore correctly puts it, 'you pay for choice'. One task of the designer, however, is to contribute a subtle and up-to-date knowledge of the aesthetic preferences in a given marketplace, so that the economic balance may be struck between uniformity and choice in the range of products being manufactured.

In the stool and the cake we saw extreme cases of the two main factors that lead to a designer being brought in to mediate between a customer, and the maker of what the customer wants. In the case of the folding stools, there was a *practical* function to be performed whose complexity was beyond the skills of customer or maker; in the case of the cake, they needed specialist help to solve a *psychological* or *aesthetic* task. In constantly differing proportions, these are the designer's two main functions. In everyday language, the designer has on one hand to make a thing that 'works', and on the other, to make a thing that 'gives off the right messages'.

With both the stools and the cake, however, we were looking at the designer's intervention in a situation that is more typical of the past than the present: the situation where one customer commissions a specially made object from a specific handicraft producer. Figure 9(C) describes this triangular relationship in diagrammatic form.

Such situations do exist still, of course, in commissions to more-or-less handicraft producers like dressmakers, fitted-furniture makers, agricultural blacksmiths or small jobbing engineers. There are also many examples of what might be called the 'modernised' version of this relationship. By this I mean cases where there is still one 'customer' who wants a single object 'made to measure', but where the technologies for making that object are no longer 'handicraft' ones. Certain kinds of machine tools for very specialised areas of manufacturing are ordered and created on a one-off basis like this. Works of civil engineering like bridges are usually also special one-off designs, as are most sizeable buildings. Inside such objects there are usually many standard, mass-produced components like bolts, gearwheels, standard steel sections, doors, lavatory pans and so on. But the composite object for which the client places the order, and pays, is largely made up of specially fabricated pieces, and the overall combination is designed to fulfil that client's unique set of requirements. This is the situation in Figure 9(D).

In general however, the problem for the later nineteenth and twentieth century designer has been different in a more fundamental way. Developed societies are now characterised by mass production techniques and vast populations of rising expectation in the middle economic groups. This has put most designers into a position which is historically quite new, of having to respond to the requirements of a generalised customer, defined only by statistics from market research. The designers in modern mass-production industries never know the person who will use what they design except as a 'typical user', or as the bundle of economic and behavioural characteristics defined by a 'customer profile'.

This change from the old craft-based relationship is of course ultimately the result of the whole basket of changes in economic and technological fields we call 'the industrial revolution'. Steam power made it possible to shape metals with enormously greater speed and efficiency than with hand tools, and production machinery for handling all other materials, natural and man-made, exploded the simple repertoire of goods previously available into whole new fields and into an economics based on high-quantity production runs.

Clearly the very fact that identical items were being mass produced immediately and inevitably broke the link between one user and one maker – a link mediated or not, as the case might have been, by a 'thinker' called the designer. We are so accustomed to today's situation where most goods pre-exist our specific personal desire to possess them, that it is difficult to imagine how different the situation was before that. Mass production, after all, is almost invariably speculative: a person with a machine produces objects in the hope of finding purchasers. There is no named person standing there to confirm that someone does actually want the item. Only statistics or other forms of information can tell the maker what this unidentified customer seeks in advance of its being made.

As you can immediately see, this is the stage at which even the designer needs an adviser: someone who does not design the solution, but designs the problem!

The more the specialised producer who is operating the machinery becomes specialised in the refinement of actual *production* (as economic pressures for efficiency dictate that he must), the more rapidly he loses contact with the user. In today's mass production process, no individual user has yet entered the situation. There is no specific family at the door requesting folding stools to fit their particular house and caravan; there is someone with information about the marketplace who reckons that demand would exist on a fairly widespread scale in certain sections of the population for a folding stool, if it could be made light, robust, easy to fold and costing no more than a certain sum.

In this new situation, two kinds of intermediary operate between producer and user.

One of these intermediaries is the 'market researcher'. The traditional market researcher's task was to study where a population's dissatisfaction with existing goods may lie, and to study trends in underlying value systems which will shape future taste or expectation. Increasingly however, this reactive position is being replaced by a more aggressive one which we would call 'proactive'. This may mean manipulation of associations and demand patterns through advertising or other less explicit media of suggestion. Or it may mean creating a whole new area of demand where none existed before, through launching of a product no-one could ever know they had 'wanted', because it didn't exist.

An example of the first kind would be the market researchers who interviewed housewives and found that what they hated most about house-cleaning was lugging around the heavy and awkward tethered-ostrich of a vacuum cleaner with body and neck all tied back to a wall-point by its electric lead. Two solutions have been designed for that problem over the years. The first, expensive and difficult to install was devised many decades ago but continues: this is a piped vacuum system installed at skirting board level throughout the house to which a neck-and-nozzle unit only need be connected wherever cleaning is needed. The other depends on new power-storing technology, and is thus less powerful, but cheaper and more flexible. This is a 'cordless' vacuum cleaner that is charged from mains electricity in a holster to become equivalent to a battery-operated machine, and can thus be used anywhere like a torch for as long as its charge may last.

An example of the second kind might be the 'Walkman' type music player. Nobody ever expressed a specific wish for it to a market researcher, because nothing remotely like it ever existed before. The Japanese company Sony which 'had the idea' themselves never expected such extraordinary success. They aimed it at a relatively limited market, but customers saw its enormous potential to create a kind of life-experience they too had never conceived. The result is one of the most successful products ever launched on the mass consumer market. Such a case depends not so much on market 'research' as on market 'creation'. This stage of the process of producing a new product is discussed in greater detail in Block 2.

Whichever process it is that generates the essential concept for a new product, somebody or some group has to convert that typical 'need' of a potential future customer into an actual object which performs that task and is capable of manufacture at appropriate cost. This second intermediary between user and maker is the modern industrial designer.

A dramatic change has taken place here in the whole relationship between makers and users of objects, and hence in the nature of design and the designer. The designer now stands in the middle as the key decision-maker, taking information from all sides, some of it based on predictions of social and economic trends, some on hunches about future psychological and aesthetic demands, some on hard science, some on knowledge of the technological possibilities available, but always in relation to a 'representative' individual in the marketplace *of the future*, not about a named individual standing on the doorstep *now*. This is the relationship shown in Figure 9(E).

Before leaving this matter of relationships it is important to clarify the terminology we are going to use to identify the various participants in this cycle surrounding design.

Back in Section 2 at the beginning of this Block we defined a 'customer' as something different from a 'client'. A *customer* essentially buys a standard or ready made item. A *client* buys a service whose precise nature or end product is decided through dialogue between buyer and seller. Manufacturers of mass produced goods therefore have 'customers', and professional people like designers, whether dealing with people as individuals or as groups, have 'clients'.

This is a general description of a type of relationship which may apply between any two participants in this process. So for perfect clarity the final purchaser at the end of the line, who was hitherto called 'the

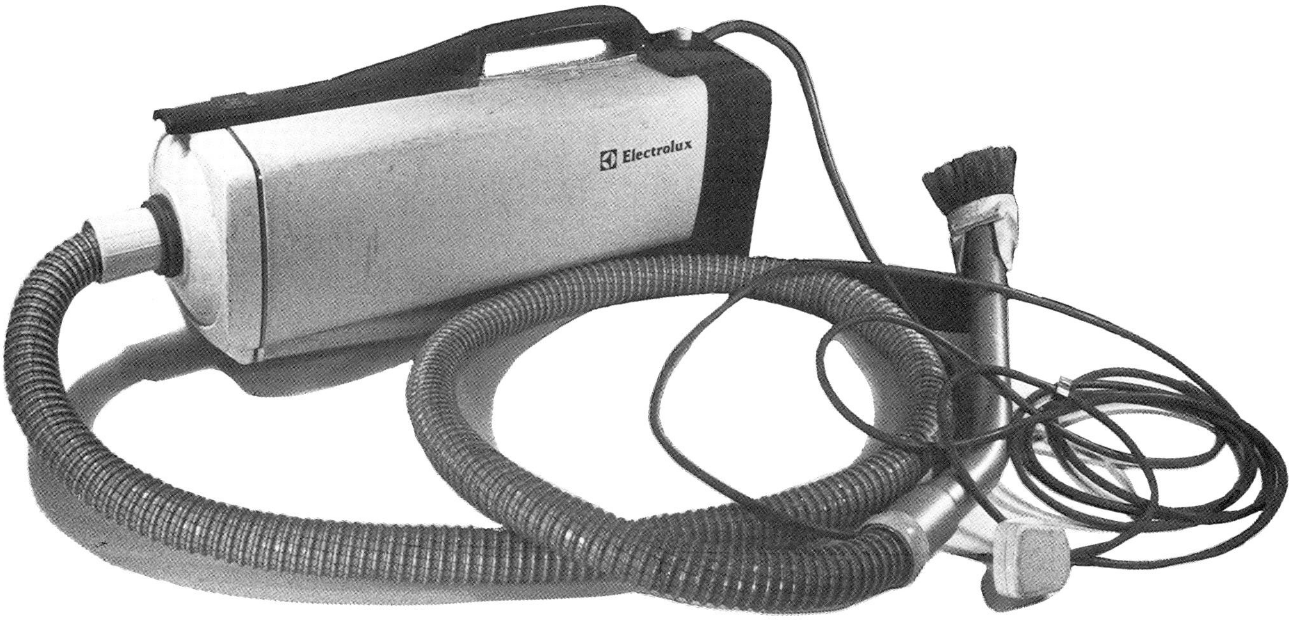

(A)

(B)

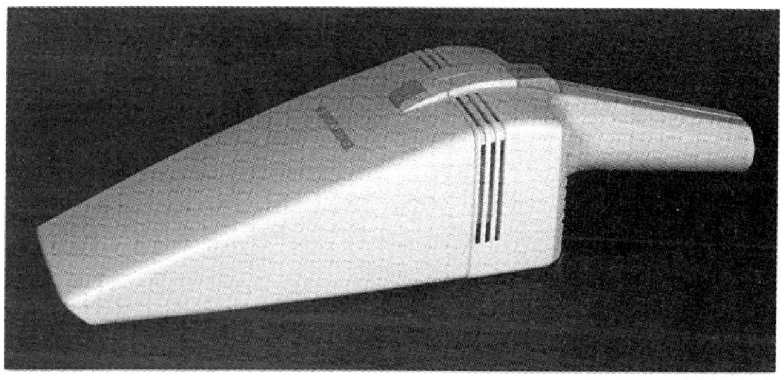

(C)

FIGURE 11

(A) TYPICAL CONVENTIONAL DOMESTIC VACUUM CLEANER, MAINS-POWERED WITH LONG SUCTION PIPE: ELECTROLUX, EARLY 1980S

(B) CENTRALISED ALTERNATIVE: THE PIPED VACUUM SYSTEM WITH SINGLE CENTRAL VACUUM UNIT AND FIXED PIPEWORK WITHIN THE BUILDING STRUCTURE WITH NUMEROUS PLUG-IN POINTS FOR THE NOZZLE UNIT: UNIVAC LTD., LATE 1980S

(C) FREE-RANGING 'CORDLESS' (RECHARGEABLE) ALTERNATIVE: SMALL DUSTING MODEL BY BLACK & DECKER LTD., LATE 1980S. FULL SIZE CLEANERS USING THIS TECHNOLOGY ARE ALSO IN PRODUCTION

customer', needs a more specific name. 'Customer' merely defines the relationship to the manufacturer or the manufacturer's agent. This customer can be both *purchaser* and *user*. On the other hand, particularly when the purchaser is an organisation, purchaser and user may be different.

When these roles split, a significant conflict of interest may arise that can potentially cause confusion of aims for the designer and manufacturer. An organisation with one set of priorities may buy the object for use by an employee who has quite different ones. Take the offices and filing cabinets we touched upon earlier. Most companies that buy filing cabinets want the cheapest equipment that is compatible with safety and functional efficiency. The secretaries who will live with them day in, day out, on the other hand, would greatly prefer the soft pink or the sunny yellow ones rather than the universal grey that is cheaper ...

Keeping the 'whole' user happy in such cases is just another set of balancing acts which the designer and manufacturer have somehow to perform as their market becomes more complex with the growing complexity of society and diversity of values within it.

In all these circumstances, however, the overall task of the designer is still exactly that which we identified with the stool or the cake: it is to determine the precise form and specification of the object which the maker should make, in all its smallest details.

It is not the designer's task to make the object in a physical sense. The designer will mentally make it: he or she will think it out. Sometimes of course the designer will also test that thinking out in actual material, experimental model or as a prototype. But the real process going on here, as the dictionary says, is one of 'making a mental plan'. This mental plan is the 'design'.

SAQ 3

Draw a simple diagram, on the lines of Figure 9, showing the participants in the case of the purple and green cake, and how they were related. Identify them by the terms defined here.

SAQ 4

Draw a similar diagram showing participants and their relationships in the case of the cordless vacuum cleaner.

7 MODELS FOR COMMUNICATING THE DESIGN TO OTHERS

Like any bit of complicated thinking, the design proposal or 'mental plan of action' needs capturing in objective form as the designer goes along. Those thoughts are only electrical pulses, and as all users of a brain know only too well, those pulses do not always make it into the long-term store! So we use forms of written record – 'hard copy' as we would say in today's computer jargon – in order to record our thinking. They do not only record the end product, they also help in processing our thoughts *en route*, to provide 'models' which help us in manipulating the bits and processes of the mental plan as we develop and refine it.

Drawing itself is no more 'designing' than typing is 'writing'. Poets can keep developing and perfecting a piece of poetry in their head and use that as its only storage, as often happens with political prisoners deprived of writing materials. Likewise the essentials of any kind of thinking can in theory be stored just in the memory of the brain. But if the maker is not also the designer, as we have seen, the problem arises of communicating that 'mental plan' to someone else. Until the moment when we can plug one brain into another, like a computer, and transmit their contents across space and time as required, we shall continue to need hard storage outside brains.

Traditionally, we make these information stores on paper. Poems and longer bits of verbal thinking are printed in books, but when the information is about objects, it is usually stored as drawings, which record the three-dimensional object in a range of two-dimensional representations. In the most advanced industrial practice today, these are tending to disappear, and the information may be abstracted even further into codes and numbers. We shall see some specific examples of this in discussion of Computer Aided Design techniques (CAD) in later Blocks.

Going back to relatively simple design problems like those stools, this description of the object could actually be a little model of a stool, in the conventional sense of the word 'model'. It could be a little miniature stool. In fact, that was how furniture makers operated in the middle ages and right through to the nineteenth century. Apprentices also made such miniature models as part of their apprenticeship examinations. Even now such miniatures of complex, one-off furniture designs sometimes turn up in antique shops. More normally though today, the 'model' which passes from designer to maker would be a drawing.

Like the miniature stool, the drawing is an abstraction and simplification of the final item, but embodies all the information which has to be transmitted from one person to another in a slightly 'coded' form. When the model is a miniature scale model, of course, the main bit of information which the maker requires to decode the model is its scale. One tenth of full size? One twentieth of full size? Confusion here will be disastrous! So between the designer and the maker there have to be some agreed 'codes' which enable them to communicate accurately. Much of what you will learn in your own drawing exercises in this course concerns the 'conventions', or standard codes, which are used to make sure the designer encodes his or her intentions accurately in the drawings, and the maker decodes the drawings correctly to produce exactly the object intended.

It may already have occurred to you that both during and after the process of designing the object, there are actually two quite different tasks of communication to be performed, because there are two different kinds of people who have to be told 'what it's like'.

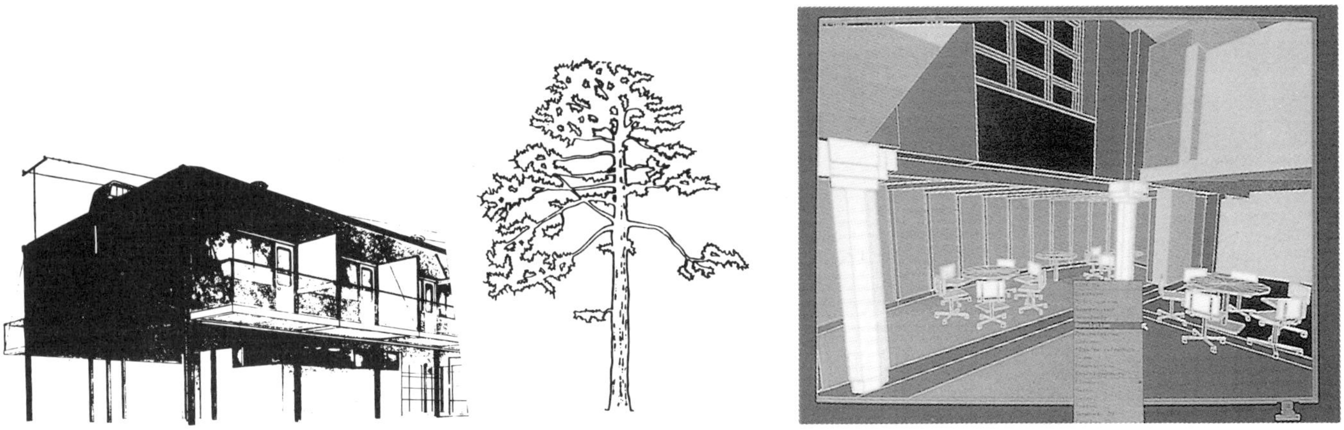

(A)

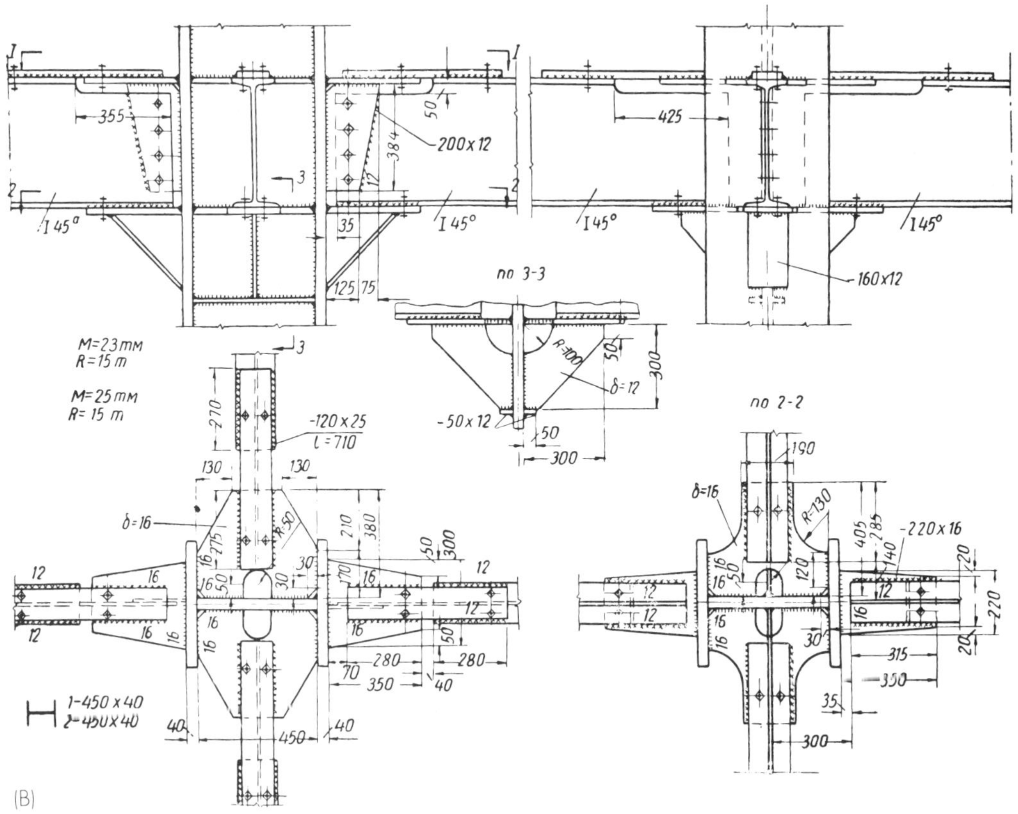

(B)

FIGURE 12
LEVELS OF ABSTRACTION AT WHICH DESIGNERS COMMUNICATE THE NATURE OF THEIR DESIGNS:

(A) EXAMPLES OF A PRESENTATION OR CLIENT DRAWING (REALISTIC PERSPECTIVE OR OTHER THREE-DIMENSIONAL PROJECTION WITH LIFELIKE DETAIL): A SMALL BUILDING DRAWN BY HAND; AN INTERIOR SPACE DRAWN BY COMPUTER

(B) EXAMPLE OF AN ENGINEERING OR WORKING DRAWING GIVING TECHNICAL INFORMATION FOR THE FABRICATOR (PARTIAL REALISM PRESERVING ACCURATE DIMENSIONS THROUGH CONSISTENT SCALE BUT ONLY IN ONE PLANE; SOME INFORMATION PRESENTED THROUGH CONVENTIONS OR NUMERICALLY): STRUCTURAL STEELWORK DETAILS

(C) EXAMPLE OF CODED OR PARAMETRIC INSTRUCTIONS (INSTRUCTIONS COMMUNICATED TO THE FABRICATOR THROUGH ENTIRELY ABSTRACT CONVENTIONS): A KNITTING PATTERN

(C)

BACK
Using 3¼mm. Needles, cast on 87 (91, 97, 103, 109, 115, 119) sts.
1st Row: K.1, *P.1, K.1, rept. from * to end.
2nd Row: P.1, * K.1, P.1, rept. from * to end.
Rept. the last 2 rows for 10 cm. (4 ins.) finishing on a 1st row.
Next Row: (Inc. Row) — Rib 7 (5, 4, 6, 6, 6, 7), inc. in next st., * rib 11 (9, 10, 14, 15, 16, 12), inc in next st., rept from * to last 7 (5, 4, 6, 6, 6, 7) sts., rib to end. 94 (100, 106, 110, 116, 122, 128) sts.
Change to 4mm. Needles.
Continue in st.st. until Back measures 58.5cm. (23 ins.) from commencement

Commence Patt
1st Row: (Right Side) K.7 (10, 13, 15, 18, 21, 24), *K.2 tog., y.fwd., K.1, y.fwd., sl.1., K.1, p.s.s.o.. *K.3, y.fwd., sl.1., K.1, p.s.s.o., K.2, rept., from * to * once more, K to end.
2nd and Alt Rows: Purl.
3rd Row: K.7 (10, 13, 15, 18, 21, 24) sts., [K.2 tog., y.fwd., K.1., y.fwd., sl.1, K.1, p.s.s.o., K.1] 3 times, K to end.
5th Row: K.7 (10, 13, 15, 18, 21, 24) *K.2 tog., y.fwd., K.1, y.fwd., sl.1., K.1, p.s.s.o., * K.2 tog., y.fwd., K.3., y.fwd., sl.1, K.1, p.s.s.o., rept. from * to * once more, K. to end.
6th Row — Purl.
These 6 rows form the patt.

On one hand the designer has to tell his *client* what the object he is proposing is like in such a way that both sides can be sure it fulfils all the specified requirements such as performance and appearance. As we have seen, this client may be an individual – and the individual in turn may be either the end-user, or the managing director, sales manager or whatever of a manufacturing company. Or the client may be a group: the members of a housing association, say, if the designer is an architect, or the Board of Directors of an airline seeking a type of passenger seat from a furniture designer, and so on. The kind of information which these people need in order to make a decision is quite different from what must then be communicated to the *maker* of the object, be it craftsman or factory, of how the thing is to be made. In traditional practice, these two types of drawings would usually have been called 'presentation' or 'client' drawings, and 'working' or 'engineering' drawings.

Since these two types of drawings have quite different purposes, you would expect them to employ quite different codes. They have to communicate quite different kinds of information, so they naturally take quite different forms.

Clients are not generally expert in reading or 'decoding' drawings that are highly abstract. Maximum realism of appearance is what they need, coupled with a clear exposition of how this object will fulfil their requirements. This kind of drawing has a job of persuasion to do. The client has to be persuaded of the merits of this object, and thus persuaded to go ahead with making it. A very similar drawing, if not a photograph, would also be required to persuade the potential customer to buy it.

The maker, whether hired craftsman or engineering shop foreman, is not involved in this strategic decision-making. In this position the question is: 'how is it envisaged this thing will be made', 'is that a rational way to do it', and then, 'can we do it here'?

At this stage, the task is one of understanding and carrying out instructions to make the folding stool, not of asking whether a folding stool is needed, and whether this is what it should look like anyway. Drawings which are instructions for making something have to take account of the process of making, both for the various individual components, and for their assembly into the complete object. The inexperienced designer will often be criticised by producers for shortcomings in this respect – sometimes, unfortunately, for designing something that simply cannot be assembled at all! He or she may have designed something that is splendid when each piece is neatly fitted inside the other, but where it is impossible ever to get those pieces inside each other in the first place! Such things easily happen, even in experienced design offices, if there is not constant attention to the process of making as well as the final object required.

This aspect of the task is called 'Design for Manufacture' or 'Design for Assembly', which are topics discussed further in Block 5. Common sense tells you this is vital – how could a designer ever be so stupid as to ignore the fact that the object has to be made? When the designer works 'in house' for a company, in close contact with a given set of production equipment and staff, a fruitful two-way dialogue can constantly be informing the designer's thought as well as challenging the production people to 'try something new'. But when professional designers are working on a consultancy basis for any client who comes their way, a considerable process of self-education may have to take place on a new job, just in order to understand the manufacturing constraints and possibilities in a particular firm or field.

The end product of all this study and thought will generally be a drawing, but this drawing needs planning and thinking out just as carefully as a piece of writing does, to be sure that it communicates the appropriate information, and communicates it clearly to the particular target audience. So the drawing itself has to be 'designed', and here as with designing the object itself there are three key questions to be asked. What is the drawing trying to communicate? To whom? And for what purpose? No designer (and that includes you) should ever pick up a pencil to make a drawing till they know the answers to all three questions!

This may sound alarming, but it is actually the very opposite. Once you understand this, you can relax, as a designer does. If a drawing is something through which the designer is merely exploring ideas for him- or her-self, it may be a very messy record of developing thoughts. Many options may have been explored and successively rejected or adopted on the same bit of paper, so that no-one but the designer can decipher the final result. But that's fine: this 'drawing' was merely a tool for thinking with. The ideas that did work in that mess can now be separated out from the duds which were part of the learning process, and recorded afresh. This new drawing will show the 'best ideas so far' – probably in a form that makes it possible for someone else to comment on them too. When some options have emerged which really are feasible, the designer (or you) already knows what the proposed object is like, and the key factor in the next, slightly smarter drawings will be 'what aspects of the object am I trying to describe to this particular person?' The client wants one set of information; a person whom you are asking to make it requires another set of information. The two kinds of drawings will be quite different, and they must be carefully planned or 'designed' to fulfil their different purposes.

In the advanced industrial scene of today, these two types of communication – from designer to client and from designer to maker – have diverged even further than before.

Until say ten years ago, both might have involved pen or pencil and paper, though of different types. Drawings to the client would have been obviously 'realistic', probably trying to show the object in an attractive three-dimensional environment. A model was often seen quite explicitly as a kind of 'toy' for the client to play with – and fall in love with – to seduce him into accepting the design proposed without further awkward analysis. Architectural models were often quite rightly criticised for that tendency. The 'working' drawings by contrast would have been a two-dimensional representation, flat and dull to the inexpert eye, showing the form of parts with exact precision of shape and size.

Today, the imaginative design office may make a computer and video simulation of the proposal, animated to show how the user will work the object, how neatly it folds into a briefcase, how easy it is to ship by sea-container – or whatever is relevant. Or there might be glossy photographs of a mock-up, showing its seductive colour or its clever way of fitting onto a wheelchair. These means of communication with the client, in short, can now draw upon an extraordinary range of technologies and media for ever more *realistic* presentation of an object's 'customer appeal features'.

Meanwhile, the means of communicating 'how to make it' are tending to become ever more *abstract*. Today, the 'design' may never be recorded and communicated to the manufacturer as a description of the object, but only as a description of the process that is required to make it. This is what happens in computer-controlled manufacture of all kinds. As far as the computer controlling the process of manufacture is concerned, the design exists only as a set of coded instructions for manoeuvres to be performed by a complex metal-working machine. If this sounds far-fetched, look back to the knitting pattern in Figure 12(C). Such patterns embody a design in coded instructions for manufacture which are just as obscure to those uninitiated in knitting technique as the computer description is to others.

Things started to move in the direction of this encoded automation with punched-tape control of production machinery. This too had a more familiar precedent, in the punched-card 'programs' used to drive Jacquard looms of the early nineteenth century. Now, in the most advanced industrial plants which integrate computer aided design with computer controlled manufacture of components, and with robotic assembly of them, the entire process is controlled by a network of computer programs. Examples of this will arise in Block 5. A description of the resulting object is not a useful piece of information in this situation, and the kind of drawing which communicated to individual machine operatives in a traditional engineering shop is largely irrelevant to such a process.

Despite all that, paper remains the most accessible, cheap and portable vehicle for communication between ordinary people, and the drawing on paper, with the accompanying detailed description or 'specification', is the medium we shall mainly use here in discussing objects, and in working out new ones in your own design projects. (Specifications will be discussed further in Block 2.) But the general principles apply as forcefully to those traditional forms of description as in the most advanced case: What are you trying to say? To whom? And for what purpose? Then you may ask, is a drawing the best or only way to communicate in this case? And even for you in the TMAs, maybe not.

SAQ 5

What are the different kinds of 'model' that a designer might use to develop a design and convey it to others?

SAQ 6

What are some of the new possibilities which computer technology has made available for this?

8 WHAT MAKES A DESIGN 'GOOD'?

Before going any further we must broach one question which is probably already arising in your mind. This is the question of what makes a design 'good'.

In specific situations, this may be a very complex matter, but the underlying principle is simple. In discussing the design of clothes, earlier in the Block, I said they demonstrated well how there is no single 'right' answer in design. There are only best or acceptable answers in relation to highly specific sets of criteria. These are social norms and conventions on one hand, and practical exigencies like climate or utilitarian functions on the other.

So the merits of a designed object are not absolute, but relative. Relative to what? Relative to what those involved see as the object's 'functions', and to the context in which they operate. As we have seen, the social or psychological functions of an object are as important to consider here as the utilitarian or practical ones. ('Psychological' and 'practical' are the two terms I have used here as a shorthand for these two categories.) If both categories are part of the function of an object, and have to be considered in the designing of it, they must also be considered in the assessment of its merits.

We have also seen that the needs of two quite distinct sets of people have to be considered in design: the human, technical and economic needs of the producers are part of this equation, as well as the needs of the users, in any full evaluation of the qualities of the ultimate design.

On top of these two, *user* and *producer*, the final object will also have been influenced by the priorities of the *designer*. However self-effacing he or she may have sought to be, there are so many small decisions to be made which are unavoidably (if subtly) valued-laden, that almost any object is shaped to some extent by a designer's priorities.

Sometimes it is appropriate, even positively desirable, that the designer's own values should be strongly expressed in the object. In fashion-conscious markets like executive office equipment, say, or where unusual technical and aesthetic treatment may be a selling point, a strong identity, like a strong housestyle for a shop or a company, fixes an item in the public mind. On the other hand, in a field like hospital equipment, where undistracted, unambiguous function for the operators is essential, the designer's 'personality' is of no interest beyond the extent to which it includes extreme attention to these priorities, and skill in resolving them.

The designer has therefore to be conscious of this and balance the priorities of all participants, including his or her own, in a manner appropriate to the case. Judgement of an object's merits will involve asking whether that balance was appropriate to the case, as well as judging individual particulars of performance.

Many different sets of requirements and constraints have therefore to be taken into account in judging the final object. So how can we evaluate whether the solution to each of them individually is good or bad? And how can we decide whether the overall solution is good or bad?

OBJECTIVE AND SUBJECTIVE CRITERIA

Where individual factors are concerned, the design's response to some of them can be evaluated quite objectively. In some cases it can even be quantified. Quantities of time, money or a given material can be measured, for example, if these are significant factors in production or use. If the object fails to perform adequately in certain entirely straightforward and practical functions, that can be established unequivocally and often with considerable precision. This is the purpose of 'performance tests', and it is an essential part of product development in any serious industrial situation. Some industries run their own testing laboratories for use by member companies: thus furniture manufacturers, for example, can have fabric and surface finishes tested for wear-and-tear or fire performance. You may already be familiar with the work done in Britain by the Consumers' Association, which runs testing laboratories for examining standard mass-produced consumer goods. The results are disseminated through their publication *Which?*, available to the whole population through individual subscription and the public libraries. The nature of this kind of testing is discussed in more detail in Block 2.

In general these exact laboratory-type tests deal with the areas we have called 'practical' function: size and ease of use for a given sector of the population; efficiency and safety of the mechanism for its task; strength and durability of its parts, and so on. As we already know, however, there are other factors and parameters to consider, the 'psychological' ones, which are matters of more subjective judgement.

In this area, success of an object may be a matter of very subtle distinctions or of responding to obscure individual preferences. Such matters are perfectly real, though they cannot usually be quantified in a way that is generally useful. In the case of a chair, 'comfort' or 'aesthetic appeal' may come into this category. This does not mean that such properties of the object cannot be analysed, discussed and evaluated. It just means that the answers will not usually be of a yes/no or quantifiable kind, measurable by a single factor, without reference to the social circumstances of use. The designer can get these factors seriously – even disastrously – wrong, for the situation concerned, although it cannot be said the solution is objectively 'bad'. The birthday cake discussed earlier, for example, would have been 'bad' in its context if the colours were those associated with last month's top-of-the-charts, rather than this month's. Other people might have liked it better though: purple and green are not what everyone finds appetising for a cake.

That case however was at least predictable: that particular colour scheme and its meaning were knowable facts. There are other circumstances where preferences are totally unknowable, and only testing on a typical sample of the intended market will show whether the object strikes some note of attractiveness, fills some latent desire or unconscious taste preference.

If 'what is wanted' is the criterion of quality in design, can a 'good' design be defined as one that is successful in the marketplace? Is the best solution to a given set of requirements and constraints the one that sells in the largest numbers?

Not quite.

IT SELLS WELL OR IT WORKS WELL

The first thing indicated by success in the marketplace is that the initial statement of the requirements did indeed correspond to a public desire. The second thing it indicates is that the designer and manufacturer responded to those requirements in a manner that the public recognised as satisfying their needs. It means you judged aright what customers were seeking, and gave it to them.

That is always good business, but it is not always proof of a good design. There are two main ways in which something that sells well may prove to be of unarguably bad design: when it manifestly fails to perform its purpose, or performs it only for an unacceptably short period of time; and where it is dangerous to the user. Again, these topics are examined further in Block 2. Here I just want to alert you to the fact that an apparently very small difference in how an object does something can make all the difference between it being safe and satisfactory, or being dangerous and troublesome. Much of design is concerned with these superficially 'small' differences of geometry and mechanical operation.

A very elegant-looking manual tin-opener came onto the British market a few years ago, its handles of then-fashionable matt black, white or orange plastic. It removed the lid by a horizontal cutting action around the side of the can, immediately below the sealed rim, whereas the traditional one makes a vertical cut round the top so as to leave the rim on the tin. Elegant looking as it was, I eventually threw it away (having been seduced by the looks initially, you see!). By removing the stiffening rim where the lid joined the sides, it left the top of a can so flexible that if one squeezed the can even slightly in order to pick it up, it distorted enough to make the soup or fruit juice overflow. It is only when the rims are gone that you realise how thin is the steel sheet from which cans are made. (As you empty one, or remove the label, you may find a series of corrugations stamped round the middle precisely to give this paper-thin cylinder some further stiffness against crushing.) That hazard of contents overflowing down one's clothes was accompanied by a razor-sharp metal edge left round the top. And instead of being left with a tin firmly held in the tin-opener after completing the cut, which could then be safely put down on a level working surface, the user had somehow to operate the opener with the can doing a bumpy pirouette on the worksurface, because only the lid, which is the bit being cut off, is in the grip of the tin-opener.

Failure in one aspect of its function might have been a price worth paying for the attractive appearance of this tool, but failure in three was totally unacceptable. Why should the simple act of opening a tin cost cut fingers and extra laundry or cleaners' bills? So for me it was back to the traditional model at a third of the price. It's *appearance* may not be 'elegant' (and eventually it rusts). But its *performance* is much more elegant.

One step up the scale in size was the similar case of the Chopper bicycle, much praised when it was introduced, a runaway commercial success with children, but heavily criticised in many quarters as unsafe.

Malfunction and hazard are the main ways in which a 'good design' can be actually 'bad' from the user's point of view. From the producer's point of view, on the other hand, the most common type of failure for a 'good' design is economic failure.

(A)

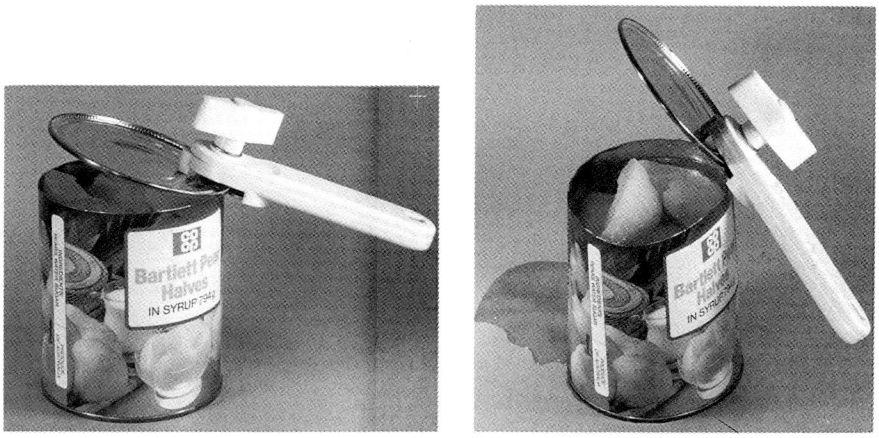

(B)

FIGURE 13
(A) TIN OPENER OF VERTICAL-CUTTING TYPE IN ACTION

(B) TIN OPENER OF HORIZONTAL-CUTTING TYPE IN ACTION

The Austin Mini, for example, designed by Alec Issigonis, revolutionised motoring in the 1960s by providing such a small and basic car. The transverse arrangement of the engine was hailed, quite rightly, as a 'breakthrough': this was the key to making the car so dramatically short. The extreme simplicity of many controls and internal features, such as the cords which replaced handle mechanisms for the doors, the windows that slid horizontally in plain metal tracks instead of winding down into solid door panels, and much else, were likewise hailed as a brilliant rethinking of conventional solutions. And they were. But even with full production lines it was not a profitable item at the prices that could be charged for such a self-consciously primitive little motor car. Only some superfluous 'added value' of trim and surface luxury, of precisely the kind that had originally been stripped out, made it possible to attract a consumer who would pay an economic price, though what they then bought had lost much of what was originally praised in the design.

In this example you can see two different approaches to the marketplace coming into conflict within the design of a single item. At their extremes, one approach aims for the simplest possible technical solution. The other creates an exaggerated sales value for what is technically and practically a bit of rubbish, by appealing to some social affectation or fashion.

In the case of the Mini, the underlying technical philosophy that inspired the design was entirely clear and consistent. Economic exigencies then forced the manufacturers to make 'social' compromises with that purity of intention in order to keep the product on the market at all.

REAL VALUE OR INVENTED VALUE; NEEDS AND WANTS; THE APPROPRIATE SOLUTION

A very extreme pair of examples which nicely polarised this argument was made famous by the American designer Victor Papanek, whom I have already mentioned as being a pioneer of the anti-elitist philosophy of design. His classic book of 1971, from which this comes, is a stimulating and provocative discussion of design for social benefit, and as relevant now as two decades ago. It is called *Design for the Real World*, and the item he calls the 'idiot gadget' (Figure 14) illustrates his larger contention, that:

> 'Much recent design has satisfied only evanescent [i.e. transient] wants and desires, while the genuine needs of man have often been neglected by the designer. The economic, psychological, spiritual, technological and intellectual needs of a human being are usually more difficult and less profitable to satisfy than the carefully engineered and manipulated "wants" inculcated by fad and fashion.' [1]

1 Victor Papanek, *Design for the Real World*, Thames & Hudson, London, 1971, pp. 10–11.

The examples speak for themselves:

FIGURE 14
HEATED FOOTSTOOL OR COOKER FROM CAR LICENCE PLATES, FROM VICTOR PAPANEK'S *DESIGN FOR THE REAL WORLD*, P. 48

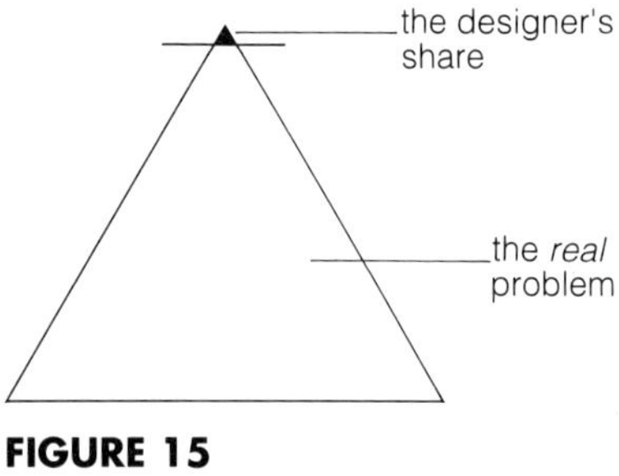

FIGURE 15
VICTOR PAPANEK'S DIAGRAM OF THE PROBLEM TERRITORY IN 'A COUNTRY', FROM HIS *DESIGN FOR THE REAL WORLD*, P. 49

As you can see from the comments here, this is a designer whose concern is with extending the territory addressed by designers from that tiny 'tip of the pyramid' which gets their services now, to 'the real problem', which in any country is the enormous bulk of its population and their material environment.

To decide whether an item is 'good', the first question will be 'where does it come on the scale between these two extremes, of the solution of a "real" need, and the solution of an artificially created "want".' Papanek is arguing that these are good and bad *morally*. At a level of global policy, that is a valid, even valuable distinction to make. But within a localised situation, in a particular factory seeking to survive economically somewhere in the West, the heated footstool may be a better way to maintain employment for a given group of workers than any attempt to make third-world cooking stoves would be.

If an item keeps people employed at the expense of richer people's susceptibility to fashionable nonsense, who is to decry the object which they so enthusiastically buy as a 'bad' design – provided it fulfils its practical and psychological function, however absurd, and is not dangerous? Is the global argument of 'a squandering of world stocks of energy and resources' more powerful than the argument from an individual community's good? A good conscience versus food in the local children's stomachs – who is to decide?

These issues are increasingly powerful within the design profession, as they are in the larger community in most Western countries. In particular cases in this course, this issue is not often addressed. This is mainly because most of our examples are inevitably already historical cases, that is, dating from before these issues became so prominent on the international public agenda. However the issue of 'appropriate technology' has several decades of history behind it already, Papanek being one of its pioneers.

As you start examining your own consumer products analytically, and puzzling over what might be 'good design' in your own efforts or amongst items in the rest of the course, here is another example of Papanek's to bear in mind. It is a powerful reminder of how much 'the economic, psychological, spiritual, technological and intellectual needs of a human being' depend on cultural context, and shows what we mean by the necessity for appropriateness.

FIGURE 16
VICTOR PAPANEK'S ACCOUNT OF HIS CANDLE-POWERED RADIO RECEIVER FOR THE THIRD WORLD, FROM HIS *DESIGN FOR THE REAL WORLD*, PP.162–165

Several years ago I was approached by representatives of the United States Army and told of their practical problems concerning parts of the world (like India) where entire populations are illiterate and living on extremely low power levels. In many cases this means that the largest percentage of the population are unaware of even so basic a fact as their living in India. As they cannot read, and as there is neither enough power for radios, nor money for batteries, they are effectively cut off from all news and communication. I designed and developed a new type of communications device.

An unusually gifted graduating student, George Seegers, did all the electronic work and built the first prototype. It is a one-transistor radio, using no batteries or current, and designed specifically for the needs of developing countries. The unit consists of a used tin can. (As illustrated in this book, a used juice can is shown, but this is no master plan to dump American 'junk' abroad: there is an abundance of used cans all over the world.) This can contains wax and a wick which will burn (just like a wind-protected candle) for about twenty-four hours. The rising heat is converted into enough energy (via thermocouple) to operate an ear-plug speaker. The radio is, of course, non-directional. This means that it receives all stations simultaneously. But in emerging countries, this is of no importance: only one broadcast (carried by relay towers placed about fifty miles apart) is carried. Assuming that one person in each village listens to a 'national news broadcast' for five minutes daily, the unit can be used for almost a year until the original paraffin wax is used up. At that time more wax, wood, paper, dried cow dung (which has been successfully used as a heat source for centuries in Asia, but for that matter anything else that burns will also work) will continue the unit in service. All the components: ear-plug speaker, hand-woven copper radial antenna, a 'earth' wire terminating in a (used) nail, tunnel-diode, thermocouple, etc., are packed in the empty upper-third of the can. The entire unit can be made for just below 9c (U.S. $ 1966).

It is, of course, much more than a 'clever little gadget'. It is a fundamental communications device for pre-literate areas of the world. After it was tested successfully in the mountains of North Carolina (an area where only *one* broadcast is easily received), the device was demonstrated to the Army. They were shocked. 'What if a Communist', they asked, 'gets to the microphone?' The question is meaningless, since the most important business before us is to make information of all kinds freely accessible to the people. After further developmental work, I gave the radio to UNESCO. UNESCO in turn would like to see it made and used in villages in Indonesia. No one, neither the designer, nor UNESCO, nor any manufacturer, makes any profit or percentages out of this device since it is manufactured as a 'cottage industry' product.

In 1966 I showed colour slides of the radio at the Hochschule fur Gestaltung, at Ulm in Germany. It was interesting to me that nearly all the professors walked out (in protest against the radio's 'ugliness' and its lack of 'formal' design), *but all the students stayed.* Of course, the radio is ugly. But there is a reason for this ugliness. It would have been simple to paint it ('grey', as the people at Ulm suggested). But painting it would have been wrong. For one thing, it would have raised the price of each unit by maybe one-twentieth of a penny each, which is a great deal of money when millions of radios are built. Secondly, and much more importantly, I feel that I have no right to make aesthetic or 'good taste' decisions that will affect millions of people in Indonesia, who are members of a different culture.

The people in Indonesia have taken to decorating their tin-can radios by pasting pieces of coloured felt or paper, pieces of glass and shells on the outside and making patterns of small holes towards the upper edge of the can. In this way it has been possible to by-pass 'good taste', design directly for the needs of the people, and 'build in' a chance for the people to make the design truly their own.

This is a new way of making design both more participatory and more responsive to people in the Third World.

Radio receiver designed for the Third World. It is made of a used juice can, and uses paraffin wax and a wick as power source. The rising heat is converted into enough energy to power this non-selective receiver. Once the wax is gone, it can be replaced by more wax, paper, dried cow dung, or anything else that will burn. Manufacturing costs, on a cottage industry basis: 9 cents. Designed by Victor Papanek and George Seeger at North Carolina State College.

The same radio as above, but decorated with coloured felt cut-outs and sea shells by a user in Indonesia. The user can embellish the tin can radio to his own taste. (Courtesy of UNESCO.)

The principles applied here, and the lessons learned, are not only relevant to the Third World. They extend, as you see, to the question of who should make aesthetic decisions. Should not individuals in society be themselves far more involved in deciding these issues by themselves, through doing more themselves? Beyond that, the question of power is raised: who exerts political control through the devices technology makes available? 'If the people have radios they may listen to the Communists' said the soldier to Papanek. When the technology is a medium of communication, it is easy to see how it can exert a political influence, but almost any object, at root, can have a more or less direct influence of generally political kind. What was Papanek's answer? Look back at the text and remind yourself of his words.

In this democratic, pluralist spirit, the question of what is 'good', as you can see, has some surprising answers. There are more aspects of this question to come.

SAQ 7

What are the two main ways in which something that sells well may all the same be an objectively 'bad' design from the user's point of view?

SAQ 8

What is the most common way for an objectively 'good' design to be unsuccessful from the maker's point of view?

SAQ 9

What is the moral dilemma pointed up by Papanek which increasingly concerns globally conscious designers now?

9 'REDUNDANCY' OR A TAILOR-MADE FIT? WHICH IS 'BETTER DESIGN'?

Papanek mentioned how he showed this little radio at the famous Higher School for Design in Ulm, Germany, back in the sixties. This school was closely associated with the product design philosophy and style developed by the Braun company. If you are not familiar with their exceptionally sleek and plain products, have a look next time you are passing a consumer appliance shop (of the more expensive kind!).

In the early sixties, just before the incident with Papanek's radio, they had established what was then quite a new aesthetic for consumer equipment. The outside form of everything, whatever its inner mechanism, became a sleek black box with nothing more on its exterior than minimal controls and their name, BRAUN, in small letters artistically placed to one side. By the mid-sixties, as the comment on 'grey' in Papanek's account reflects, the appearance of their products was relieved occasionally – in those items like food mixers not readily reducible to 'boxes' – by a non-cubic form and by dark or two-tone greys.

At that time, transistors and other electronic devices were just beginning to enter consumer goods. This treatment of their appearances was an expression of the notion that advanced technology was not something acceptable only in industrial or commercial environments, but would henceforth become a natural part of the domestic, even intimate, environment of civilised people.

A quarter of a century later, that does not sound a very revolutionary thought, precisely because their prediction was factually correct. By the early nineties, very ordinary kitchens in our culture are filled with sophisticated and specialised food preparation equipment. When a car with central computer control of its engine performance is backed out of the ordinary suburban family's garage today, wood and metal-working machinery equivalent in power and precision to the local carpentry shop of a whole village a few years ago, may be set up to perform some relatively simple DIY tasks. That is quite apart from the almost absurdly high fidelity of domestic entertainment systems piped to all rooms inside quite ordinary houses, and home computers that can access worldwide data systems from the living room (most of France already gets its phone directories in this electronic form, for example).

One of the most difficult and important questions which this explosion of mass accessibility to technology has produced for the design policy maker, and hence the designer, is that of appropriate levels of 'fit' or 'redundancy' in design. In simple terms, this means how much strength or capacity should the object have, over and above that necessary for its immediate and explicitly 'intended' tasks?

Of course this is not a new question, but it has acquired a new complexity in circumstances where complex and powerful machines are being made available to massive sections of the population who have no special training in their use, and will use them in ways and situations entirely beyond their manufacturer's (or anyone's) control. It has become a legal minefield of liabilities, trades descriptions and definitions of 'fitness for purpose' or 'reasonable use'. But our interest here is in the implications for the design approach. These implications can be summed up in two questions. First, how far should an object be capable of doing other things, or more of the same, beyond its declared capabilities? And second, how far should it be required to be safe in the hands of untrained, unskilled and unsupervised users?

The answer to both questions can be 'just enough', which I have called 'a tailor-made fit', or it can be 'more than enough', which means some agreed level of 'redundancy', i.e. a level of power or capacity over and above the 'normal' requirements which it is publicly declared to fulfil. As soon as anyone starts designing anything, it becomes necessary to adopt a clear policy on both these questions, for almost every element and component and for the object as a whole. 'Is the level of redundancy appropriate?' is a key question in deciding 'is this a good design?'

What has this got to do with Braun? I will give you a pair of comparative examples from the early period of this domestic equipment we now take for granted, which points up this issue well. As on some previous occasions, I give you a personal example. It may reassure you that even designers learn the hard way with their own purchases. But more importantly, it will encourage you to be thoughtful in your evaluation, and honest in your criticism, of the industrial products you yourself use and buy, for this is the first step to understanding the thinking and decision-making processes which caused them to have particular characteristics. Indeed, the first step is to realise that they do not have particular characteristics through chance, but through someone's conscious decision (or neglect of a decision). And these decisions could always have been otherwise.

Call this, if you wish, 'the true parable of the hair-dryer and the hand-held electric whisk'.

We go back to the early sixties, before there was clearcut policy and legislation on such matters, when this was even more than today a matter of particular design philosophies.

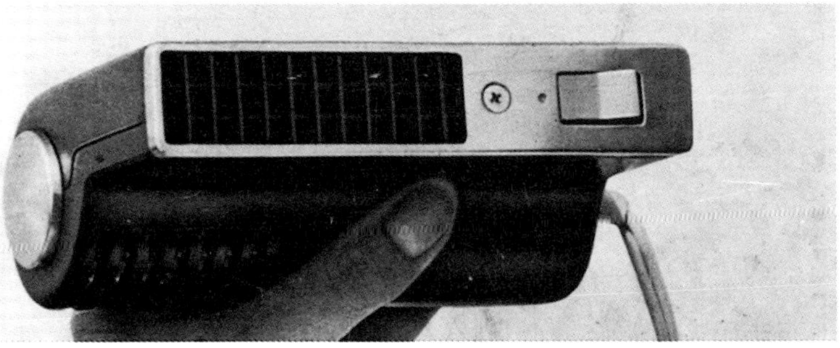

FIGURE 17
BRAUN, COMPACT HAIRDRYER FOR USE 'AT HOME AND WHEN TRAVELLING', 'EVERYWHERE AND ALWAYS TO HAND', DESIGNER REINHOLD WEISS, 1963

Braun produced a hair-dryer, the first of the compact ones that presented itself as a personal accoutrement rather than an extension of a whole family's household equipment. It was of course matt black, and shaped like a box with rounded corners, the boxy bit becoming a cylinder to accommodate (and 'express') the rotating fan element at the back. It was not a lot bigger than the palm of one's hand; it had all the stylistic austerity I described as their particular innovation, as well as being of changeable voltage for worldwide use. Its case actually was a pure matt black box, slightly soft and padded. Its price was a fortune to a young designer, but I bought one, explicitly in order to have the experience of living with such a model of 'the best modern design'.

It did the hair all right, though more with a sense of austere efficiency, as was then becoming fashionable, than of the sensual pleasures of 'having a bath and washing one's hair' to which today's curvy little machines (or the earlier 'household' ones) responded. In today's market, that item would be sold as a man's hair-dryer: efficient personal technology for the appearance-conscious modern male, discreet alongside the laptop computer in the briefcase. But in those days men almost never used hair-dryers, very few had a personal one. Design on the other hand, was predominantly a male world by comparison to today, and had ruthless deadlines. Equipped with this tool not available to my colleagues, I started to use it for drying the thick ink lines which were the worst time-consuming factor of architectural drawings in those days. It looked just right in that context: the aesthetic and the shape invited one to see it as an extension of 'professional equipment'. But it was a disaster. After being turned on and off quite frequently during one all-night drawing session, it packed up. It did not have any overload cut-out: something just burnt out. An irate battle with the manufacturer's British agents ensued, whose upshot was a statement informing me that this machine was designed for relatively short periods of use – half an hour or so – a few times a week, and never for frequent on-off use in intensive shifts. I had given it years of use in a few days, and could not expect compensation when a product failed through such gross misapplication.

I was of course furious that an item promising this new world of broad technological power was actually so limited in its capacities. But I learned an expensive lesson in the apparent callousness with which manufacturers could construct a contradiction between image and reality. This apparently all-powerful little hot-air machine had no redundancy at all: it was very literally and only 'a hair dryer'.

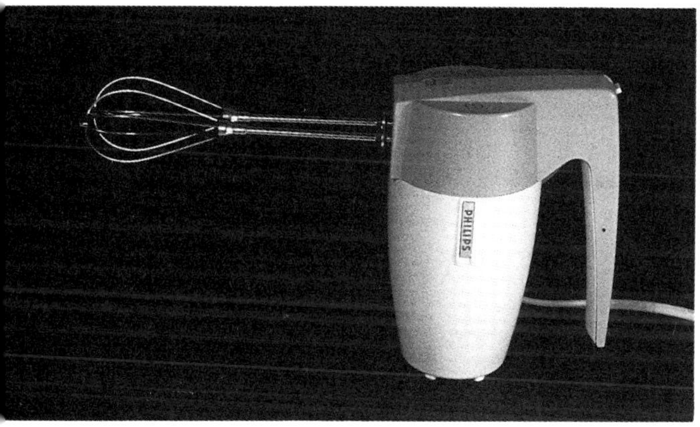

FIGURE 18
PHILIPS, HAND-HELD KITCHEN MIXER, C. 1960

Contrast the philosophy of the hand-held electric food whisk, of similar date. This too was long before the explosion of kitchen machinery in the seventies and eighties. At that time, these new products were being launched into a relatively innocent public, but in this case that innocence and its practical implications for possible users of the machine were taken into account as a positive element of design policy.

In an extraordinarily extravagant gesture towards 'saving domestic labour', as it then seemed, I purchased a hand-held food mixer made by Philips. It was a quiet, unassertive object in beige and grey, with pleasant curves where appropriate: no macho image here, just a friendly tool. It was sold for whipping egg whites, making a traditional sponge, stirring a sauce or a soup – that kind of light but arm-aching chore. It was a novel item at the time, and when I chanced to meet a senior executive of Philips I expressed appreciation of several of its features (it was particularly easy to clean and well balanced in use). I asked what its real capacity was, and he laughed. 'You can be sure we wouldn't launch such a product unless it was actually capable of mixing an English Christmas pudding,' he said, 'because some people will inevitably use it for that, whatever we say on the box, and complain if it doesn't do it!' Thirty years later, it has nobly stirred many weighty Christmas puddings, and is still going strong.

You see the difference here, and the dilemma. What should be considered the normal and reasonable territory of application of a product when users' imaginations will inevitably find many other applications beyond that for which it is explicitly designed and sold? Indeed, is that 'functional redundancy', that capacity to do other things as well, to be considered a feature of 'good design'?

The answer, of course, is that sometimes it is and sometimes it is not. There are occasions when a tool or piece of equipment is very properly designed to perform a limited and very specific task in the best possible manner. This is what I have called 'a tailor-made fit'. The correct point on the scale between 'tailor-made fit' and 'extensive functional redundancy' depends heavily on the anticipated levels of skill and technical understanding in the target population. In more general terms, this is the anticipated balance between their imaginations and their prior conditioning. Their imaginations will see all manner of expansive possibilities, and their prior conditioning will limit them, either by exerting a sensible caution, or by causing them to be disciplined by standard conventions as to 'what the thing is for'.

The ability to explode those conventions, to see the underlying nature of a device behind its conventional label, is one of the key skills a designer must develop if precedent is to be any use as a source of inspiration in problem solving. Some of the creativity techniques discussed in Block 3 are at root ways to get through 'labels' to learn from the common principles embodied in artefacts we have conventionally allocated to quite different 'functions'. Before we leave this aspect of the fit-versus-redundancy issue, here is another parable – this time a genuine (i.e. fictional) one which forces us to think about this question by imagining a provocative extreme case.

The tale was told to the international design conference at Aspen, California, in 1955, by the American designer Bernard S. Benson, to make a similar point about 'what is "good" design'.

> 'An island inhabited by natives is suddenly blessed with a bevy of washing machines; electricity too, of course. There are washing machines in abundance but no manufacturer's operating manuals, nor in fact any statement as to what the machines are for. Over a period of time, one tribe comes to the conclusion that the machine is for kneading dough, makes very fine bread in fact, and they are very happy. Another tribe decides that the machine was created for separating the corn from the husks and is very well satisfied. Another group decides that the purpose of the machine is to keep the babies of the tribe happy while the mothers work. The babies are put on the inside, of course, and the mothers stay on the outside. (The machines are slowed down for the purpose.) Mixing paint and making pea soup are the obvious purposes of the machine to other tribes, and everybody in his conviction is very happy. Nobody, incidentally, has thought of using it for washing clothes.
>
> 'An Operating Manual ... did not come with the product, and as a result the customers in different groups and at different times evolved their own.'

As he concludes:

> 'Because of this wide diversity of human objectives and goals, it is difficult to establish criteria for the evaluation of any creative activity [i.e., in this case design – CC]. Although one cannot find a common denominator at the cause level [i.e. in functional characteristics of the object itself], it seems one can find a common denominator at the effect level. Here there is universal accord that it is desirable to achieve "happiness", although the paths that lead to this effect vary tremendously among people and at different times, and may often be diametrically opposed.' [2]

That conclusion forces us to recognise that 'redundancy' also applies in the other area of 'function' we have identified: psychological function. Given the objective is 'to make the user happy', an object whose aesthetic, as well as its practical abilities, can please many people, or please the same person in many different moods and circumstances, will have wider success than one which only evokes or supports a very limited range of psychological responses. If you think about the implications of this you can see one reason why aggressively 'specific' or 'character-ful' objects – like the Braun hair-

2 Bernard S. Benson, 'Leisure', in: R. Banham (ed.), *The Aspen Papers, twenty years of design theory from the International Conference in Aspen*, Pall Mall Press, London & New York, 1974, pp. 37-38

dryer, for example – are a risky product for the manufacturer seeking a mass market.

The other area in which redundancy has to be considered, as I mentioned earlier, is the area of safety, which includes reliability. This is simpler, more susceptible to quantified prescriptions, and easier to understand.

In this category we have the argument which says that motor cars should have engines considerably more powerful than is necessary for driving up to the speed limit of 70 m.p.h., because that extra power reserve provides safety to 'get out of scrapes' in an emergency. We have tables quite strong enough to stand on, because they may be used for that, although intended only to carry the plates of a meal. We have building regulations which decree that walls, floor and roof members must have certain reserves of safety over and above their 'design loading', and so on.

Housing offers a rich example here, as it provides interesting contrasts in both areas of 'fit' and 'redundancy' – contrasts to which much of the population responds in a very clearly calculated way when choosing property, without always recognising the principles they are following.

Old houses – Victorian terrace houses for example – were built with great reserves of safety. By comparison to a standard speculatively built house of the end of this century, such houses have great technical redundancy. Walls are far thicker than is needed to hold up the roof, and holes can be knocked in them with relative impunity: indeed whole walls removed in many places. Likewise roof joists readily carry the extra load of an attic room conversion, and so on. Compare today, when all these elements have been reduced to the bare minimum, and one realises why old houses are so popular amongst those wanting flexibility or a personalised home. But it extends beyond the extra capacity in physical loadings. Rooms in such houses, and the arrangement of the rooms, offer possibilities for a great range of different uses. This is partly a matter of sheer size: the tiny bedrooms in today's unregulated housing design, where no minima can be enforced, are such small 'people cupboards' that they offer opportunities for little more than sleeping. But it is also a result of how their space is organised. When every space is accessible to the circulation system (i.e. hall and landing), for example, you have a much more flexible configuration than in an open-plan arrangement, for converting the building to quite another use – such as multiple occupancy, or even offices. The effect that spatial organisation has on the properties of a design is something examined in detail in Block 4.

Finally we have the simplest version of 'redundancy for safety', which is reliability. This is basically the principle of the 'belt and braces'. If the priority of a given design is extreme simplicity and lightness, say, there may only be one element capable of carrying each particular loading or executing every function. But there are many cases in which safety and the possibility to 'keep it going' mean that overlap and duplication are essential features of a 'good design' for that situation.

SAQ 10

Write down in your own words the problem lying behind the issue of 'fit and redundancy'.

SAQ 11

Give two reasons in favour of including some 'functional redundancy' in objects for general consumer use, and two reasons why it may not be a good idea.

10 WHAT'S IT ALL BEEN ABOUT ?

From everything that has been said already, you will see that we cannot even begin to talk of 'designing', i.e. thinking of an object, until we have done a lot of thinking about what that object is for, about what in the broadest sense are our aims and intentions, and about how we understand the link between the object and the people who use it.

Before we explore this a little further with a specific type of object, let's just sum up the main principles we have established so far.

1 Designing is conceiving a plan for doing something, or usually in our case, for making something.

2 In any situation where a 'problem' needs 'solving', the answer may be a combination of material and non-material interventions, a combination of 'designing an object' with some element of change in organisation, management or communication. Objects are just part of a larger continuum here.

3 Even when it is clear that a new object is required, the 'function' of an object is only partially utilitarian or practical. In differing degrees, it will always have an element of psychological or even symbolic function too, defined by the social aspirations and cultural environment of its user. At the same time, it has a different but equally complex function, which at root is economic, in the life of its producer.

4 In different situations, the designer has different relationships to the various other participants in the process, in particular to the producer or maker of the object, and to its eventual user. In all cases, the designer's job consists of assembling a very diverse range of information, part of it objective 'fact', part of it dependent on the priorities or 'values' of different participants, and synthesising this into proposals for an object which will answer the needs which have been identified.

5 Representation of the object, in a drawing or other medium, is not the essence of designing itself, but is the designer's essential tool, both in thinking out the actual design idea, and in communicating the features of what he or she proposes to the people who will use the object, and to those who must make it.

6 Judgements on whether a design is 'good or bad' can never be made by abstract criteria without reference to the full range of functions (practical, psychological, economic, etc.) it has to fulfil for all participants in the design and making of the object, and in its life in use thereafter. The weighting of these factors will vary with the balance of priorities in each case, and may differ at different times in the life-cycle of the object. This multi-variate quality is integral to the nature of design: many variables have to be analysed, given values and evaluated simultaneously. An object may be 'good' in relation to some variables, and 'bad' in relation to others. Complex questions of determining 'appropriate' or 'reasonable' levels of performance also arise in relation to these criteria, and must be decided within the overall design process.

PART 2

LET'S SIT DOWN FOR A WHILE

11 WHY ARE ALL THESE CHAIRS SO DIFFERENT?

We have had a brief encounter with some stools. I'm now going to use their big sister, the chair, as an example through which to explore some of the factors influencing design in more detail. In particular we shall see how the motivations of different kinds of people lead to the initiative of launching a new product onto a market, and how different materials influence the forms of the results. When it comes to the TMA, we shall come back to many of these examples and case-histories to extract options and raw material for your own design exercise.

The chair is a very common object in our culture, though not in all cultures. In different variants it illustrates a wide range of design problems and approaches. It opens up many of those aspects of design we have already identified, in both practical and psychological categories.

'Sitting on something' is not, for a start, a uniform process. There is an enormous number of typical and standard sitting situations in our sedentary culture. Each of these involves the body being held in a characteristic posture. Those postures are subject to anatomical limitations, of course, but are basically determined by social conventions: by what is considered polite or impolite, formal or informal, what is considered appropriate to 'authority' and to 'obedience' and so on. If chair and posture do not fit the conventions of a given social situation, a person feels 'wrong' or even embarrassed in a group, and the whole 'event' may be distorted, socially awkward or unsuccessful as a result. I'm sure you have known many such occasions: when everyone else is relaxing in deep armchairs at a party, you feel very awkward as the one sitting on an upright chair brought in from the dining room. Conversely if a formal job interview with someone senior in the company is conducted in those same low reclining chairs, it can be very difficult to make one's points seem businesslike and to respond as crisply as if one was sitting at a table.

We all also have experience of chairs that are badly designed in objective respects, regardless of the social context to which they are appropriate. Their dimensions may be wrong and cut into the legs behind the knees, or fail to support the back adequately; they may be structurally unsound; they may have rough joints or protruding elements which tear at clothes or skin.

Cast your mind's eye around your own home for a minute, and think of what chairs it contains.

Now flick through the pages of the rest of the Block, to get an impression of the types of chairs we shall examine here.

Some of them may have a generic similarity to ones in your own home, some will probably not. So why have I chosen this particular range of examples?

There are a number of reasons, related of course to what we are seeking to learn from them.

First, they are chosen as examples which were innovative. In many cases they were in some way a 'first': the first to apply a certain technique or material, for example, or to solve a particular problem. All are generally regarded as good design of their type, and are famous for that reason. As you will see, I do not always support the view that they are in fact 'good', and shall explain why in relevant cases.

Second, with a few contrasting exceptions, they are items one would call typically 'modern'. In a technical sense, this means that the materials of which they are made are treated as an important design factor to be expressed with honesty, or exploited imaginatively, in their own right. This interest in materials brings these selected chairs closer to other, more obviously industrial kinds of artefact we shall examine elsewhere in the course. In this respect, the deeply padded armchairs, mentioned above, are different. The structure which supports the body weight is concealed – only prior experience leads us to suppose they probably have a wooden frame, for example. If relatively new, the comfort element is probably provided by rubber webbing and foam upholstery, if older, by metal springs and horsehair or feather cushions. But we cannot see that just by looking at the chair. For contrast, our last two examples will be of this kind and as you will see, the motivations for them, and the messages they try to convey, are of a different character from those underlying the other chairs.

Many of these chairs are characteristically 'modern' in a social sense, too: they provide for sitting in public places of types characteristic of the modern world. So they are designed for mass cafés and restaurants, for reception areas to amenities and offices, for multi-purpose spaces which at times need to be cleared by stacking chairs on top of each other, or in one case, for the mass transport by sea which exploded in the pre-war years before mass air travel. These are new classes of use which result from the situation of a democratic mass market for services which has developed in the West during the late nineteenth and twentieth centuries. Again, for contrast, we have some chairs which have nothing to do with that 'modern' world.

The **third** factor governing my choice was that virtually all the chairs are the work of a named designer. In many cases, these are famous names from the history of design in general, or of modern architecture. In that sense they serve as a little introduction to that whole world of professional thinking about modern artefacts. But my main reason was a desire to stress the fact that everything is designed by someone. Or it may be designed by some collective of several people, and as in the proverbial case of the camel, that 'horse designed by a committee', the result may be something of a compromise. But here too, the result can only be evaluated if one knows what the conflict of priorities was within that committee.

In some consumer-goods fields, and in some places, there is a 'cult of the designer' which means that every object gets a label with a name on it. In industrial fields, very famous names of proven performers may be seen as 'guarantees' of some kind. But most of the artefacts we handle are anonymous in this sense. This is why it is so easy to forget that, for better or worse, they could have been designed differently. The good qualities of an object which performs well may be chance, but they are far more likely to be the result of good design work. Conversely it is easy to be complacent about the thousands of everyday objects which perform badly, which could have been easier to use, easier to hold, easier to open and so on, if only a better *design* had been worked out by the people concerned before they started *production*. Such complacency amongst consumers or users is the biggest disincentive to the improvements of design which can do so much for individuals' lives, our national culture and simply for our collective commercial prosperity.

Thus, if an object has a designer's name attached, one is reminded there is someone to be thanked or complained to, for its characteristics. Most things don't just 'happen'. In the slow evolutionary processes which we discussed earlier, we have lost the names of the people who made the successive incremental improvements, but it was precisely those people's design talents which made the resulting traditional objects so satisfactory.

My **fourth** and final criterion in selecting these examples was a slightly patriotic one. We are accustomed now to a situation where design of British consumer goods is not always particularly good, and where design is not seen as something of official concern except in as far as it may help international sales of British goods by individual companies. Within living memory, however (at least for slightly older people), there was a quite different attitude at government level. I refer to the period during and immediately after the Second World War. Maybe the attitude we shall see then was paternalistic, and maybe such a concern is inappropriate to the different circumstances of almost half a century later. But that period offers an important example of another set of public values, which is important to understanding the possible priorities of design. So approximately half of this international selection are British examples, and there is a cluster of examples (forming Section 5) which trace lines of thought and development in the decade from 1942 to 1952.

What is a chair? 'A separate seat for one, of various forms', says an Oxford dictionary. The chair differs from the stool in having a back support, the stool being defined as a 'backless seat for one'.

What interests us here are those 'various forms'.

Why are they all so different?

If chairs are one of the most essential and common items in our culture, what are we to conclude from the diversity of forms they take? That no-one has yet solved the problem of 'the separate seat for one'? Or that everyone is always playing around with making them different 'just for the sake of being different'? If it's the latter, are they trying to give chairs that 'built-in obsolescence' which is so profitable to manufacturers of anything, by making them into a fashion item, or have designers taken them over as a kind of 'art' item through which to do a little personal expression?

There are elements of both those factors to be found amongst the reasons for diversity. Surprisingly, both have positive aspects. Being 'different for the sake of it, to capture new customers', is another way of describing that drive to find better solutions which is the main motor of most industrial innovation. Not infrequently it is an urge 'just to be different' which generates the drive to find a way of applying a new material, plastics, for example. Likewise, the designer 'expressing a personal philosophy' will probably be expressing values or aspirations that are widely shared in that society, and those of like mind will be pleased to welcome into their environment something that 'speaks to them' of their own values and concerns.

In our survey we shall come upon examples of these two motivations for yet again redesigning the chair, but there are of course many others.

Despite the fact that they are more or less generically 'chairs', in that they provide a sitting posture which (in most cases) supports the back, they are all in fact solutions to rather different problems. As we shall see, there was quite a different set of priorities in each case.

In the first group, the problem was to find the **absolutely minimal structure** that can give the body's muscles physical rest.

In the second, the sitting positions being accommodated are so **specific to a given culture** that the resulting objects could hardly be used in any other.

In the third, the chair is **pure symbol**.

In the fourth group, we see responses to particular situations in the **availability of different materials on the market**.

The fifth group shows how quite different forms arise through designers **exploiting different properties of the same basic material**.

In the sixth and last group, we see two 'seats for one' where both shape and materials have been chosen purely **to evoke sensual associations**.

Good design, as we have seen, is not an absolute. In the words of that apt traditional phrase, it is a matter of 'horses for courses': a horse that is of good physique for running a short, intensive race on the flat is different from the horse which is good for a long rough steeplechase – and both are different again from the horse which is good for pulling a plough. The qualities which make a good horse for each case are specific, and well understood by experts, but they are different. The same applies to these four-legged artefacts, as to all others.

12 THE PURE AND MINIMAL STRUCTURE TO GIVE THE BODY PHYSICAL REST

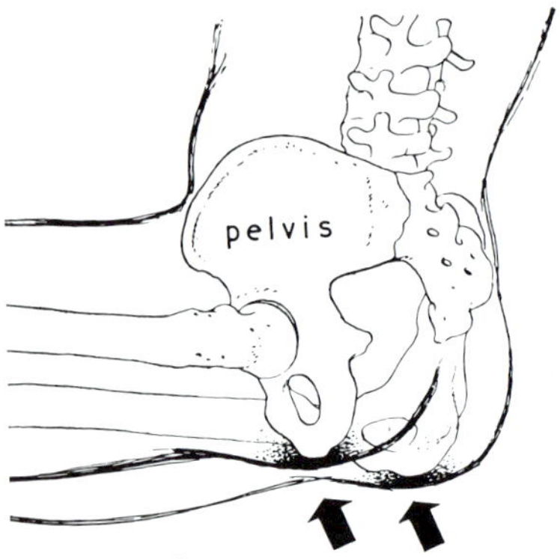

FIGURE 19
WEIGHT-BEARING POINTS OF THE PELVIC BONE IN CORRECT SITTING POSITION, FROM LOGIE, 1947

A: VERTICAL SUPPORT – THE COMPRESSION STRUT

The essence of sitting is propping the body in a position where most of its muscles can relax, without the person falling over. The body's weight is taken off the legs, and transferred up to the pelvis.

The simplest independent structure which can take that weight is a plain compression strut slightly shorter than the individual's legs, on the butt end of which he or she can rest the pelvis as if on a third inanimate leg. By spreading the legs apart, a tripod is formed in which the person is stable, without wobbling, and most of the body weight is carried to the ground by the strut. The end of the strut needs to be wide enough for both sides of the pelvic bone to rest on it, otherwise it will be hideously uncomfortable after more than a few seconds. Given some little T-shaped bar across the top, such a minimum strut is comfortable. You may have noticed that bus stops and underground railway stations have lately gone over to little 'bars' like this in place of proper seats, to save space and to stop tramps sleeping on benches, they say. But their little bars are hinged off the wall or the bus shelter's frame: you cannot carry them around with you. To be able to sit where you choose, you need the independent strut as well.

The application of this principle of the minimal compression strut for sitting is perfectly demonstrated in the traditional item called a 'shooting stick'. If you have always lived in a town, you may never have seen such a thing, but to country people they are regular traditional equipment. They were originally devised, as the name indicates, for people going shooting. In order to steady himself (it usually was 'him' in the days when these were invented), the man aiming upwards to shoot a bird would lean back sitting on this 'third leg', forming a stable tripod with the bottom half of his body that left him free from the waist up to twist or bend as he followed the bird's flight in his gun-sights and shot it down.

In the advertisement reproduced as Figure 20, from about 1900, you see several slight variants on what this dealer calls the 'shooting seat'. The strut was usually steel and its fittings brass, but Mr Mills, as you can see bottom left, keen for lightness, is already using aluminium in what he calls his 'patent sportsman's stools'. At twice the price of those above, these are also 'telescopic': like a photographer's tripod, the height can be varied. But flashing that aluminium around as a status symbol, as you may notice, is counter-productive to the sport. Under the prices (centre bottom), the dealer advises: 'We strongly recommend the Black Bronzed Stools for Sportsmen, as the Bright [i.e. polished aluminium] may be too conspicuous on a sunny day and frighten the birds.' Another important aspect of the practical function is being carefully considered!

The seat always folds, as you see, forming a handle. No.8A here, top right, has two little pads of 'plaited cane' within the frame of the seat-cum-handle. This seat part, usually about 10 cm wide and 20 long, was commonly formed of a simple leather sling, though that was slightly less good in really wet weather. At the bottom of each strut you will have noticed a little disc. You probably guessed what that's for. Its function is to spread the man's weight over an area of ground and stops the tip simply disappearing into the soft earth under his weight – and him falling over on top of it. At the same time, the spike is essential to fix the position and stop the disc slipping sideways.

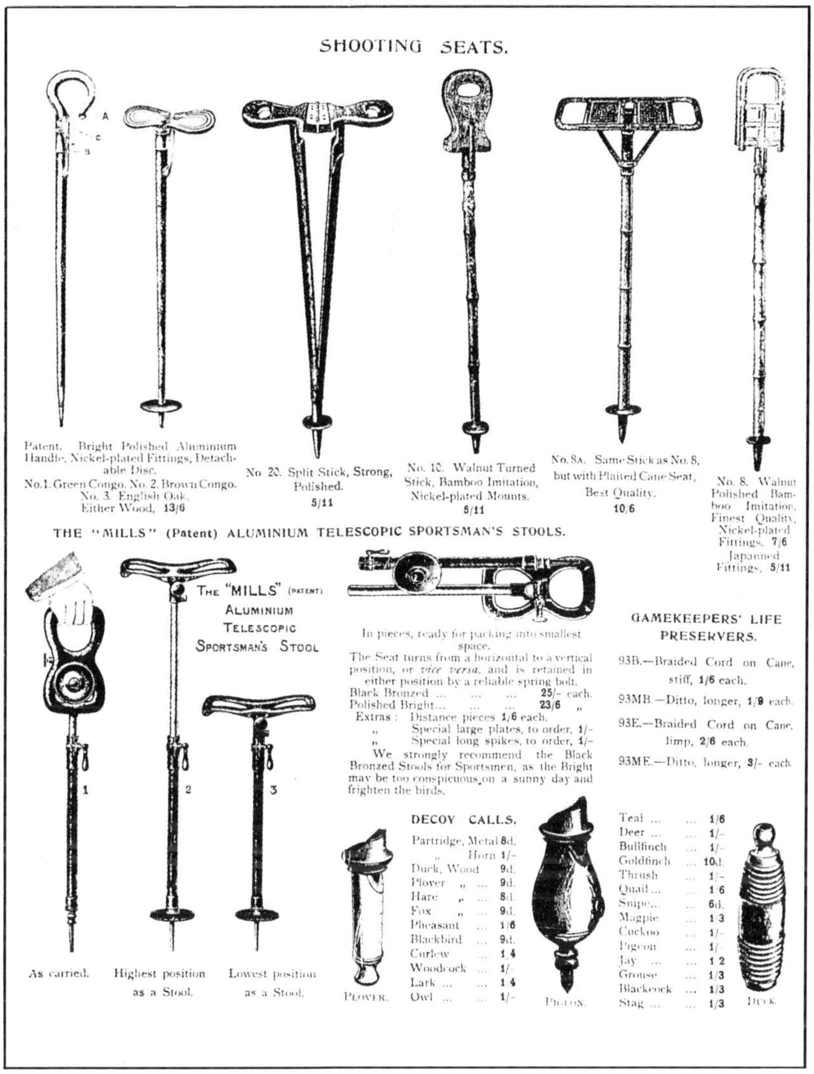

FIGURE 20
ADVERTISEMENT FOR SHOOTING STICKS FROM THE CATALOGUE OF WILLIAM WHITELEY LTD., LONDON, EARLY 1900S

Quite apart from their original use amongst farmers and others shooting birds, these shooting sticks are useful for anyone walking in the country (you see how they also function as a walking stick when the seat is closed), or for people watching sports in open sports fields. As far as I can discover, no particular 'inventor' or 'designer' is credited with devising the first one. This is therefore an example of something that has evolved anonymously in numerous hands till it is genuinely a 'classic': it is a simple, efficient solution to a problem that has evolved to a form which serves all aspects of its practical purpose so well, and is so satisfying to use, that it does not need to change further.

You may be wanting to protest that Mr Mills was right in calling these 'stools', as they have no back. You would be right of course: these are not strictly speaking 'chairs'. I have included them as the purest answer to one half of the task performed by the chair: the task of taking the weight off the legs by supporting the body in the *vertical* dimension. The function of the chair's characteristic back support is to prevent the body falling over in this position: it provides support in the *horizontal* dimension.

Our second example in this category of 'pure minimal structures' complements the first. Where the shooting stick only gave support in the vertical dimension, this only gives support in the horizontal.

FIGURE 21
SITTING POSITIONS, FROM THE CLASSIC SWEDISH STUDY BY BENGT AKERBLOM, *STANDING AND SITTING POSTURE*, STOCKHOLM, 1948

B: HORIZONTAL SUPPORT – THE TENSION STRAP

On a stool without back support, muscles of the back have to maintain tension to prevent the upper body falling, and the most important point of support is the lower back, the so-called lumbar region. If this is supported, the whole spine is kept straight and upright above it without further effort. At normal sitting height this is serious, but if a person is sitting on something very low, or worse still on the ground, even greater muscular effort is required to keep a right angle at the lumbar region, rather than let the spine curve backwards from there unsupported. If it does sag backwards, muscles start to ache with the tension, and in prolonged situations, the spine itself may suffer permanent injury.

If there is nothing solid against which this part of the back can be leaned, how are the muscles to be given this horizontal support? It is a problem to which both improvised solutions and commercially manufactured products have long responded in the same way: by a form of sling that supports the lumbar region of the back with a tight loop passing forward around the knees. In this structure, the tendency of the spine to sag backwards is resisted by the tension in the strap. As the sling tightens, the tension is transmitted to the upper leg, which becomes a compression strut.

One such example in current commercial production comes from the United States, and calls itself the 'Nada-Chair Back Sling'. The makers describe it very aptly as 'a legless chair that allows perfect posture without strain'. Boating publications have pointed out that 'Because of its frameless design, the chair can be used ... to provide back support where none is offered, such as in canoes and fishing boats'. Walkers and campers endorsed it with the observation that 'Nada-Chair ... allows you to sit comfortably with ample back support anywhere', as the illustrations show, even on flat ground or on a beach. The materials are nylon fabric and padding, webbing and adjustable buckles – all light and washable.

FIGURE 22
NADA-CHAIR BACK SLING, MID 1980S, IN USE IN A SMALL BOAT AND IN CAMPING

Like our first example, this too is one specific example of a general class or type of object. It is not so well established as the shooting stick, though in present day conditions it ought to be far more widespread than the former, since very high levels of crippling lower-back trouble result from unsupported sitting by people who live sedentary working lives and are muscularly unfit.

In our analysis, as the manufacturer's own phrase 'a legless chair' indicates, this device complements the shooting stick, which is nothing but a leg, in exemplifying the application of pure principles of chair function through the simplest possible means. They are responses to quite different situations: you won't need much prompting to think of two strong reasons why a shooting stick is not the solution to sitting in a small fishing boat, for example!

Both of these objects are designed for outdoor leisure and sporting situations, and for people moving around. They are intensely functional, and do not have any extraneous social or symbolic role to fulfil. Both have to be minimal structures: as light and portable as possible. Given these priorities, in one case the problem has been analysed down to a recognition that it is really only a vertical support which is required; in the other, vertical support at an appropriate level can be improvised from the environment, and the vital function of the device is horizontal support.

The essence of a true chair, as the dictionary definitions made clear, is that it provides both. In less exigent fixed environments in and around buildings, it can afford the amount of material and solidity of structure to do so.

13 TOOLS FOR THE BODY TO SIT IN HIGHLY CULTURE-SPECIFIC POSTURES

A: BODY TOOL NO.3, ERIK DE GRAEFF, 1975

The ease with which European and American populations have welcomed Japanese consumer products into their lives in recent years has tended to obscure the enormous differences between the cultural environments of producers and customers in this case.

Those in European industry who must compete directly with Japanese goods in the marketplace are generally only too aware of how significantly certain deeply rooted aspects of Japanese behaviour patterns and social codes have contributed to that industrial success. In relation to the chair, however, the difference is very significant.

FIGURE 23
KO-SHOIN INTERIOR, KATSURA IMPERIAL PALACE, KYOTO, JAPAN

Japanese domestic and social life is generally so bounded by formality and inequalities of role between the sexes as to be intolerable to most Westerners for long periods. At the same time, the positive side of deep adherence to traditional values is a calm and repose of domestic spaces, and a role for symbolic distinctions, that has never been characteristic of individualistic European culture. Thus many people have found there (as in other contemplative, nature-oriented Eastern philosophies), elements of life style whose calm spirituality they see as an important corrective and balancing factor to the ever more frenetically object-obsessed and self-obsessed tendency in Western cultures.

The traditional conception of a Japanese home is not of a place of activity, clutter and hobbies like an English home. It is a form of shrine, peacefully filled with spiritual symbolism about man's relationship to nature. Very simple, abstractly composed and proportioned spaces are its essentials, with calm horizontal emphasis under a low ceiling, and as for chairs – in a traditional Japanese home they do not exist.

People sit on the floor, on mats or cushions. So strong is the connection between sitting and the act of being together, that anyone who passes through the room without sitting down does not 'enter the room' socially at all. They need only be acknowledged as 'present' when they come down to the level at which the room is 'inhabited', that is when they sit down.

Through lifetimes of training and use of appropriate postures, this low-level sitting does not produce those back problems it does for us, which we discussed in relation to the 'leg-less chair' above. To a Westerner who seeks for philosophical reasons to live in a somewhat Japanese style of simplicity, however, the problem is a serious impediment. De Graeff's aim was to overcome this problem with an item which in itself incorporated the Japanese philosophical values through applying the design principles which characterised traditional Japanese artefacts.

(A)

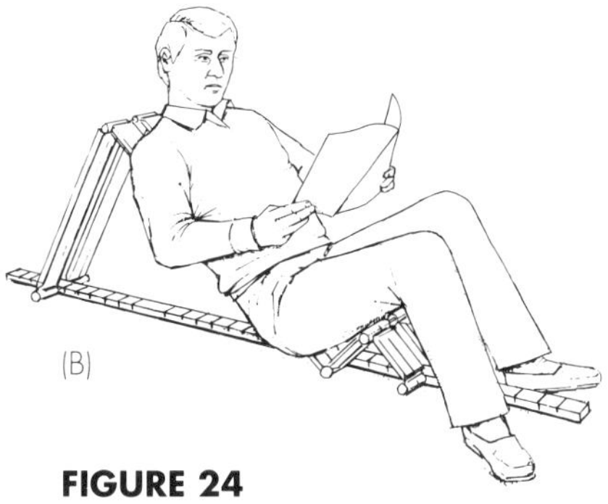

(B)

FIGURE 24

(A) ERIK DE GRAEFF, BODY TOOL NO.3, 1975: PHOTOGRAPH

(B) SKETCH SHOWING DE GRAEFF'S BODY TOOL IN USE

He calls his sitting structure a 'body tool' because it 'must exist in harmony with the body as might a good tool'. If it is to perform in harmony with all bodies, it 'must be infinitely adjustable for all bodies', that is for sizes and angles of sitting. Unlike a Western room, where the chairs remain there permanently with an independent identity and presence, creating a 'social grouping' even when no people are present, Japanese spaces are empty when the people leave them. Thus the sitting tool should not take up space in use, and should fold away into a minimal volume when not required by a person. In accordance with basic Japanese principles of design, 'the object should be reduced to a perfect, simple and flexible form.' [3]

Those are de Graeff's own phrases. Let us look at this more concretely with specific examples.

First, the positive philosophical importance of all that empty space.

> 'One salient characteristic of Japanese art and culture is the emphasis on *ma*, a space or interval. In Nō plays, for example, a small hand drum is struck: the beat is followed by silence; then the drum is struck again; that beat is followed by silence; and so on. This *ma*, the interval between the soundings of the drum, which is characteristic of Japanese music, is not related to the pause or rest found in Western music. In Japanese music, even when there is no sound, something vital remains.
>
> 'To the Japanese, *ma* is not just an empty space, where nothing exists: it is the space or interval necessary to give shape to the whole. One of the more derogatory terms in Japanese is *manuke*. In use, its meaning is something on the order of "grossly stupid", but its literal meaning is simply space (*ma*) that is missing (*nuke*). That "*manuke*" is considered such a strong insult that it can lead to blows is, I think, a good indication of just how important the presence of space, or ma, is in Japan.' [4]

That comes from a book on carpentry details called *The Art of Japanese Joinery*, by Kiyosi Seike. You will see why I am stressing this, when we come to compare an English example shortly.

3 Erik de Graeff, quoted in: Sharon Plant, 'The Evolution of the Modern Chair', *SIT'89*, exhibition catalogue and display notes, Business Design Centre, London, 1989

4 Kiyosi Seike, *The Art of Japanese Joinery*, Weatherhill/Tankosha, New York & Tokyo, 1977, p. 16

FIGURE 25
TRADITIONAL JAPANESE JOINERY: THE OKKAKE-DAISEN-TSUGI RABBETED SCARF JOINT

De Graeff's words 'perfect' and 'simple' have very clear implications in this context. You can see in the typical Japanese interior how the whole is made up by assembling very clearly defined components in a very simple and logical way, so that none of the components is cut into or otherwise interfered with by another. Each fulfils its own function calmly and independently: to screen, to be walked on, to let light through, to hold the roof up, and so on. We call this 'clear articulation of the parts'. The beautiful geometry of traditional joints in Japanese joinery have the same clear articulation where each part does its specific job, and you can see exactly the same characteristic in de Graeff's sitting tool.

Like the traditional Japanese house, it is made of natural materials, the dominant one structurally being timber. To a Japanese, this already has some immediate implications for the geometrical form. As Seike says in the same book, 'In wooden architecture [in Japan], the majority of construction employs straight lines, primarily because of the characteristics of wood and the nature of its growth.' [5]

The long base is plain wood, carrying the little cross-bar ridges onto which the body support slots for the different positions. The paired timber elements which support the sitter's back and pelvis are covered in natural canvas. The hinge joints are simple steel elements. Each piece in the overall linkage performs its function clearly: the base is about 8 cm wide, and each of the supporting members is about 10, so the load-bearing pelvis has a comfortable width of 30 cm or so on which to sit. As with the shooting stick, the sitter stabilises the posture by placing the feet somewhat apart to make a triangle on the floor. The whole 'tool' is painted calm middle grey.

Even the word 'tool' here has some richer connotations than would be conveyed by any English equivalent. It is the nearest possible English word, but in the context of carpentry, 'tools' mean something philosophically much more complex to the traditional Japanese craftsman than they do to us. Seike again gives us a very clear description of this.

> 'Although modern technology has given us new materials and new tools, Japanese carpenters still use a very special, respectful term for their instruments, even their electric tools. No Japanese carpenter refers to his instruments as mere "tools", but instead calls them *dōgu*, which really has no equivalent in another language but means roughly the "instruments of the Way [of Carpentry]." *Dōgu* is a very old word and perhaps it has not disappeared because even though materials and tools have changed and improved, carpentry methods and the "Way of Carpentry" have remained basically unchanged for hundreds of years. ...

5 Seike, *Japanese Joinery*, p. 17

> 'Originally a *dōgu* shop was one dealing in implements and equipment for the tea ceremony. The various implements used in the tea ceremony – *sadō*, or the Way of Tea – were called *dōgu*, or "instruments of the Way [of Tea]". Carpenters tools also came to be called *dōgu* because for carpenters there is a Carpentry *Dō*, or Way of Carpentry, and carpenters too considered their implements *dōgu*, or instruments of their Way. ... Perhaps we could even say that formerly the carpenter's *dōgu* were invested with a degree of divinity.' [6]

So the concept of what the 'tool' is for, in this Japanese context, is not purely functional. De Graeff's phrase the 'body tool' has connotations of being 'an instrument of the Way'. It is something whose role in the individual's life is to contribute on a path of discovery, a spiritual or philosophical journey. The implication, as you will have grasped even from what we have seen here of a traditional Japanese building, is that 'truth' is found by emptying the conscious mind and the senses of the accidental and the material, by an act of supreme distillation that produces the 'positive emptiness' of *ma*. As we saw from his initial intentions, de Graeff's Japanese inspired Body Tool was designed to embody all that cultural background. Without an understanding of all that, it is impossible to understand why this particular 'seat for one' takes the form it does.

B: GENTLEMAN'S LIBRARY READING CHAIR, THOMAS SHERATON, 1802

FIGURE 26
THOMAS SHERATON, GENTLEMAN'S LIBRARY READING CHAIR, 1802

Look back at de Graeff's design, and look at our next one, and some pretty powerful differences must strike you immediately. 'Economy of the straight natural lines of timber', 'clear articulation of parts and how they join together': not much of that here. To be called 'a body tool', it 'must exist in harmony with the body as might a good tool', says de Graeff. What justification is there here for applying his term to this early 19th century English chair, which apart from anything, seems to be half back-to-front?

A tool is judged principally in relation to the task it enables a person to perform.

So what is the task here? For a man to sit in his library (a place, incidentally, that was male territory, rarely entered by females), and read a book with good illumination of the pages, adjustable height, placing and angle of the book itself, which is supported in an ideal position for his personal needs, and a position for himself which is well supported not only vertically, but in the horizontal, in a manner appropriate to poring over the printed page. This item is aimed at a household of roughly the same social class as the Japanese house we saw, that is the class of non-labouring, rich people who had the time to pursue and develop the traditions of intellectual thought, philosophical enquiry, and general cultural activity, handed down by their forebears, and to do it in some peace and comfort.

For this, the famous furniture designer Sheraton has produced a perfect piece of practical equipment. The reason it appears half back-to-front is that it was indeed sat on facing 'the back', legs astride the specially narrowed 'neck'. This is where it is significant that the Library in grand houses was not women's territory: this posture would not have been so readily acceptable in the drawing room!

6 Seike, *Japanese Joinery*, p. 18

Not only could the angle of the book be adjusted to choice, by the strut and ratchet device visible underneath it, but the whole bookstand could move around the half-circular armrest element on a little rack and pinion gear, operated by winding a handle. With elbows nicely supported at near-shoulder level on the two 'arms', and the book placed exactly to his preference, the reader could continue in the greatest comfort till the candles burned out, very likely moving only his right forearm occasionally to reach the glass of brandy on a side table of similar height.

So, in practical terms, this is a pretty good tool for the body. But what larger indications does it give us of the difference between how these two cultures conceive 'the Way'? Compare the two interiors here, both in their own terms the cultural centre of the palatial houses concerned (and to roughly the same scale). On the right is the Music Room in the same Katsura Palace in Kyoto where we saw another room earlier. On the left, the library in a typical English country house, not at all the grandest: this is from the fine Queen Anne style mansion called Anthony House, built in the 1720s by an unknown architect for the Carew family in Cornwall.

On the left, every surface of the library and its anteroom is busy saying something, from the classical panelling marking out the shapes of the wall and the archway to the realistic portraits, which never allow the inhabitants to forget the physical appearances and practical deeds of their forebears (the chap on the left had been famously killed by the Roundheads in the English Civil War, 80 years before this house was constructed). There is nothing very spiritual about these ancestors' presence: they inhabit the house as lifesize persons still in the flesh. In a desperate agoraphobia, furniture with oddments of no special aesthetic merit on it stands everywhere.

As the repository of wisdom in the house, the library tells us that truth and enlightenment are achieved by filling the conscious brain with the maximum amount of useful, real-world knowledge. The Japanese room

FIGURE 27
VIEW INTO THE LIBRARY, ANTHONY HOUSE, CORNWALL, BUILT 1720S

FIGURE 28
MUSICAL INSTRUMENTS ROOM, KATSURA IMPERIAL PALACE, KYOTO, JAPAN

declares the opposite: that truth and enlightenment are achieved by *emptying* the conscious mind of real-world knowledge. To the busy positivists of the panelled English library, reassuringly filled with more knowledge than one person can ever absorb, the austere 'emptiness' of the Katsura rooms is intensely threatening. Here the rich aristocrat's soul, like the poor monk's, is bared. There is nothing of the real world to distract the individual from contemplation of the eternal questions, indeed he is inescapably obligated to search for 'the Way'.

Sheraton's nifty mechanisms for the 'reading tool' were an early example of a trend in furniture that became increasingly strong through the later nineteenth century, in England as in America. People who could afford it loved the convergence of furniture with machinery. Such elaborate attention to solving the comfort requirements of an individual increasingly replaced the purely social function of sitting. At the same time, such specialised mechanisms made chairs into items of less universal application. Not only the cost of the item itself, but the affluence of space needed for numerous functionally specialised sitting places, made them the privilege of those with money and large houses.

Through close study of these two very different variants of the 'separate seat for one' you have seen something of the enormous loading of cultural values that such an object can carry. I deliberately chose designs which illustrate the greatest possible contrast, but no artefact is free of such a loading. Only good designs, however, whether attributed or anonymous, will generally distil cultural values to such purity or communicate them so lucidly as these do.

Our next chair, too, carries an enormous loading of cultural meaning, but has acquired it, and communicates it, not through any characteristics given it by a designer. It has acquired its meaning through long historical association, and through a common agreement by a particular group that the object itself, as a relic from their common history, shall be sanctified as a symbol within their social structure.

14 THE CHAIR AS PURE SYMBOL

THE CORONATION CHAIR, 1300

FIGURE 29
THE CORONATION CHAIR, 1300

Here we have an object which by the normal criteria of design has no merit whatever, and potentially only the interest of its extreme antiquity. (Virtually no other piece of English wooden furniture from that date remains.) Its shape disobeys all the basic rules for the sitting posture, with flat seat, vertical back and front. It has been heavily vandalised. But the British Monarchy and hence the State depends on it for its legitimacy. In particular, their power to rule Scotland depends on the presence of a piece of rock remaining stuck in the bottom of it, and on the rock being there when the Monarch is crowned in this chair.

The object and its history were well described by the design historian John Gloag in his book *The Englishman's Chair.*

> 'The design has been attributed to Edward I's goldsmith, Adam, and the account for the year 1300 ... suggests that the original intention was to cast the chair in bronze; but the idea was abandoned, and an oak chair made instead, at a cost of 100 shillings, without apparently any alteration to a design that had been contrived for metal. ...
>
> 'The chair combined the functions of state seat and war memorial. Described in the original bill as "a chair for the Scottish stone by the altar and in front of the feretrum", it was designed to display the Stone of Scone, not only as a trophy of Edward I's military prowess, but as a sacred object with a mystical and magical history. It was the "Stone of Destiny"; legends accumulated about its origin, and by the late sixteenth century it was confidently identified as the stone "upon which the patriarch Jacob reposed when he saw the angels ascending and descending a ladder reaching to heaven", as Frederick, Duke of Wirtemberg duly recorded when he visited Westminster Abbey in 1592. ... Originally the chair was impressively magnificent, and from the wardrobe accounts of 1300-1 it appears to have been the actual work of Master Walter, the "king's painter" to Edward I. The four lions on which the chair rests were much later additions, possibly Elizabethan or Jacobean.
>
> 'What we see today is a despoiled and mutilated wreck of a chair that formerly blazed with gold and colour; for there are indications that the whole surface was covered originally with flat gilded gesso, whereon patterns were incised by dotted lines. The turrets and leopards that once surmounted the pinnacles at the back were removed early in the nineteenth century, and for generations destructive nonentities were allowed to carve their initials and names on the woodwork. Nobody at that time seemed to have any respect or sense of responsibility for such survivals of medieval craftsmanship.' [7]

If you go to Westminster Abbey now, you will see this last is a major understatement. The chair is carved with graffiti like an old school desk, and indeed the blame lies mainly with Westminster School boys several centuries ago, left to roam in the Abbey with idle hands.

As Gloag's description indicates, there is no logical relationship between form and material here: it was designed for one material and made in another. All aesthetic and intrinsic value of the original gilded covering

7 John Gloag, *The Englishman's Chair: the origins, design and social history of seat furniture in England*, Allen & Unwin, London, 1964, pp. 35-36

FIGURE 30
CROWNING OF H.M QUEEN ELIZABETH II IN THE CORONATION CHAIR, WESTMINSTER ABBEY, JUNE 1953

has gone long ago. So have parts of the original form, being replaced by others of no special aesthetic merit whose date is unknown. Yet this is a literally priceless item on which the Monarch's legitimacy substantially depends. You may recall the constitutional crisis which threatened in 1950, when fanatical Scottish Nationalists removed the Stone of Scone from the base of the chair on Christmas Eve and whisked it away to a secret location in Scotland. A tremor went through the whole nation, accustomed then to stability and more than now bewitched by the mystique of such objects. King George VI was sick, though not yet ill – what if there should need to be a Coronation?

Constitutional crisis in the Union was averted, as the Scottish Nationalists returned the stone in April, and there was no need for a Coronation till 1953. So few are the objects which our Protestant society has imbued with real mystique, however, and so matter-of-fact have we become, that the deep human desire for the 'sacred' and 'magical' is almost outside our experience and imaginations. Yet that is precisely the myth-making, symbol-making trait which has given power to so many 'seats' in history.

15 DESIGN REFLECTS THE IMMEDIATE TECHNOLOGICAL CONTEXT: WARTIME AND AFTER IN BRITAIN AND AMERICA

A: WARTIME CONSTRAINT: UTILITY DINING CHAIR NO.3C, EDWIN CLINCH, FOR THE BOARD OF TRADE, 1942

During the 1920s and 1930s, as examples in Section 16 will show, there was considerable experimentation in the use of non-traditional materials and techniques in the European furniture industry. Though Britain was less involved than Europe with the new design movement that generated such experiment, British designers and manufacturers too experimented with the use of steel tube, pressed steel sheet, and laminated timber forms as well as numerous, by then almost traditional, forms of bent-wood.

The further development of this field, as of so many others, was enormously distorted by the Second World War. When we look at them analytically, the various design responses which emerged at different stages of the War and the recovery period offer a barometer of the changing problems and potentials of industry in different times and on different fronts.

Hovering as we do now between the totality of nuclear war and the regional skirmishes which touch us only as news items, it is difficult to reconstruct a picture of how much of life and industry continued in this country, and under what constraints, during the war from 1939 to 1945. Were people still making anything so inessential to the military effort as furniture, you may ask?

Initially there was some furniture production, up to strictly limited quotas. But in late 1940, so many people lost everything in the serious aerial bombings of the Blitz that their capacity to remain useful workers in the war effort as a whole was threatened by complete domestic disorder. Second hand furniture prices rocketed beyond the means of those in need.

With maritime supply lines cut, timber, being imported, was extremely scarce, but so were all other materials. Even plastics were restricted entirely to the war effort, as were most metals. So scarce was aluminium, as the lightest industrial metal, that a government campaign had already collected in most of the nation's aluminium saucepans as scrap for the building of Spitfire aeroplanes. For all the pre-War experimentation, timber was still the most familiar material to the British furniture industry; thus in response to the crisis, the Ministry of Supply announced in January 1941 that some timber supplies would be released to make 'domestic furniture of a simple type' for evacuated families and those who had lost their homes through bombing.[8]

That November it was announced that only twenty essential items could be made at all, and the Ministry laid down the maximum amount of timber to be used in each. No-one would be able to buy such furniture without an official 'Certificate of Need'. What started as 'Standard Emergency Furniture' became part of the larger programme of government approved 'Utility' items permitted for wartime production. This was a programme of maximum overall value for the money, resources and manpower invested by the nation at large, and quality was as strictly defined and controlled as were quantities of material and costings.

8 Timber Controller, Ministry of Supply, quoted in: Richard Sword, *Utility Furniture and Fashion 1941-1951*, exhibition catalogue, Geffrye Museum, London, 1974, p. 9

FIGURE 31
EDWIN CLINCH, DINING CHAIR MODEL 3C, PRICE £1 AND 9 SHILLINGS, FROM THE OFFICIAL CATALOGUE OF UTILITY FURNITURE ISSUED IN 1943 BY THE BOARD OF TRADE. THE CHAIRS WERE OF OAK, WITH 'PADDED SEATS COVERED WITH LEATHER CLOTH IN A VARIETY OF COLOURS'

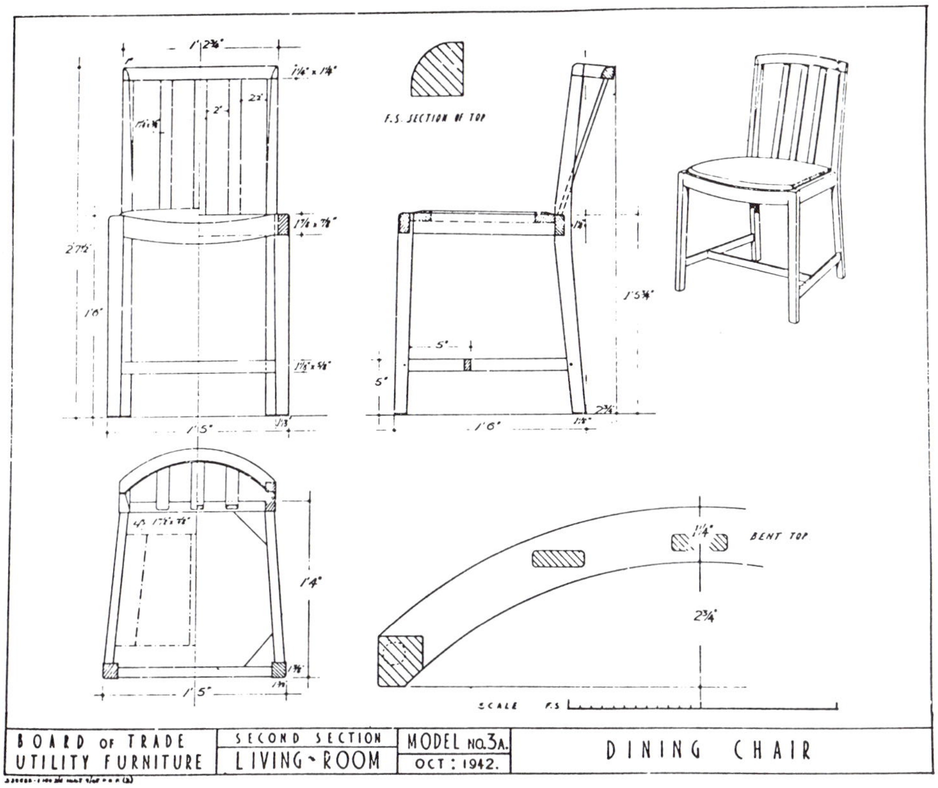

FIGURE 32
WORKING DRAWING FOR UTILITY DINING CHAIR MODEL 3A, BOARD OF TRADE, OCTOBER 1942

The Board of Trade got in 'some of the best available designers of furniture' and was also exploring possibilities for producing temporary furniture of paper and basket work. The eventual designs, produced in mid 1942, were described by one of the designers Edwin Clinch as an exercise in 'plain common sense'. 'The more simple the furniture the more precise must be the workmanship' he said, and 'All that you see with utility furniture is pure, it's good'.[9] Comparing the absolute simplicity of their range with the ornate and flimsy short-life items of pre-War popular taste, furniture, said Clinch, must never go 'back to what it was like before the War'.

To the *Architect's Journal*, whose memories contained different images of pre-War taste, from the elegant curvilinear experiments with the new materials by the design avant-garde, the new stuff was 'very ugly to look at'. But public and popular press hailed its 'serviceability' and low prices.[10]

These concerns were of course closer to the priorities of designers and the government, as the design client. The philosophy was sound materials in the minimum effective quantities, combined with the best possible standards of manufacture. Thus the materials were hardwoods, but cut, as you can see, in the simplest, therefore most economical cross-sections and here, as in most of the other Utility furniture designs, used in straight lengths. Steel screws were permitted by the Ministry, as providing a vast increase in strength despite the scarcity of the material. Seats were of comfortable shape, padded in leathercloth.

They were not to the taste of professional design fashion makers in the architectural press, maybe, but the technical principles of economy in materials and manufacture, and the social principle of quality goods for poorer people's money, were exactly those which had inspired much pre-War design experimentation by the avant-garde in Europe, when the new

9 Edwin Clinch, quoted in: Sword, *Utility Furniture*, p. 13

10 Quoted in: Sword, *Utility Furniture*, p. 14

materials had still been readily available. This is a typical example of how the same underlying principles applied in different circumstances produce a quite different kind of object. As timber chairs, these Utility designs have long been recognised as a classic solution to providing the maximum product for the minimum cost and materials under the particular combination of conditions facing client, designer, manufacturer and ultimate user in the middle of a major war. They looked austere, but because dimensions and angles were right, they were comfortable for a wide range of people and situations.

B: POST-WAR LIBERATION: BA ALUMINIUM DINING CHAIR, ERNEST RACE, 1945

Take a look at our next chair, and its date, and compare with the last one. Plainly something has changed here. Designed in Britain in 1945, just as the war was ending, it cannot have been a luxury product. Yet tapered legs look an expensive option when we heard in the case of the Utility chair that economy depended on simple uniform sections of material throughout. All those curves, too – and why doesn't it need bottom struts between the legs? The reason for all this is simple. This chair is of course made of a totally different material. The 'most economical shapes' are therefore also quite different in this case.

FIGURE 33
ERNEST RACE, BA ALUMINIUM DINING CHAIR, 1945

These parts are cast from aluminium, a light-weight metal, not cut from solid timber. The tapering is part of the economy. Structurally, a strut of this kind needs only to be fat in the middle, where it will tend to bend under compression loading from the top. If you could prevent all the sideways movement that tends to make it snap off from the seat, it could be tapered top and bottom. But that gets very elaborate. Better to thicken up the top to make a solid bracket for fixing rigidly to the horizontal seat element, but the bottom can still taper. (The principle is just the same with a stiletto heel, for example!)

Whereas it costs more to turn a bit of timber down to a tapered form, because of the labour and energy involved, and it only produces useless scrap material in shavings or sawdust, in casting a metal it does not cost any more to make moulds for a tapered form than it does for a straight one. And the molten metal you don't need in one mould is poured straight into the next. The immediate result is at least twice as many legs – possibly more – for the same amount of metal, through studying its properties and the method of shaping it.

Along with all that metal saving, the taper gives a light and frisky air, very noticeable if you compare to the previous chair again, which admirably reflects the sense of a load lifted, of an easier future, that naturally infected the whole country with the ending of six years of war.

So technology, economics and aesthetics went hand in hand here, in the typical manner of a good design.

But I can hear you protesting about all this aluminium: wasn't aluminium so scarce that people were being asked to hand in their saucepans? Yes, but that was five years earlier. When the war ended, there were thousands of tons of aluminium scrap available from the dismantling of precisely those aeroplanes to which the saucepans had already contributed. It was going dirt cheap, as no-one really knew what to do with it.

Furniture was not the obvious answer which sprang to everyone's mind, as aluminium had been used relatively little in this field before the war. Its lack of stiffness did not suit the 'cantilever' form so fashionable amongst designers, as we shall see later, which was ideal for springy steel. There had been some good stacking chairs for outdoor use produced in Switzerland on the eve of the war, which combined tubular aluminium legs with a seat and back unit pressed from aluminium sheet. Apart from anything, the 'feel' of pure aluminium, though warmer to the touch than steel, could become a little rough and unpleasant without coating. Most important of all though, in Britain, was a general conservatism that had prevailed in the popular furniture market before the war.

The Utility programme, like the war in general, had changed a lot of attitudes. When new British industrial design was exhibited in 1946 at a national exhibition called 'Britain Can Make It', the pre-War pioneer of good mass furniture Gordon Russell wrote thus:

> 'The Utility furniture scheme has been a great social experiment. It has at least given the furniture trade time to take stock and think. And it has proved that people are very ready to accept a better standard of design. Its workmanship, too, is, on the whole, better than at least 75 per cent of the furniture produced before the war. I am one of those who believe that this upgrading may be permanent. I believe that we are again reaching the main stream of sane British furniture making, which we have left for too long. I believe that today we are more likely than we have been for a century to accept the astonishing developments which are proceeding in materials and techniques. And by this I mean to use these developments thoughtfully, unselfconsciously and honestly to solve our new living problems. In talking to many of the younger people coming back from the forces I get the impression that the age of make-believe is over. Think, for one moment, what a revolution this means. It will be possible to solve definite problems of storage, sitting, table-space, writing space and so on by the best means and using the most suitable materials. It won't always have to "look like Chippendale" or have a period label. It will be permissible to use solid wood, plywood, plastics, metal, what you will, not because Sir Walter Raleigh always did, but because the material chosen is the best one for the job.' [11]

Two of those 'younger people coming back from the forces' the previous year had been an architect turned industrial designer named Ernest Race, and a production engineer from manufacturing industry named Noel Jordan. Both in their early thirties, Jordan had spent the war in the Navy and Race in the Air Force. Race knew all about those thousands of tons of aluminium scrap going cheap from the planes, where in some models all fittings including seats, as well as the skin and the main airframes, had used this metal. Race designed, Jordan understood manufacture, and more than a quarter of a million of these chairs were produced from 850 tons of

11 Gordon Russell, 'Furniture', in: W. H. Newman, ed, *Design '46: A Survey of British Industrial Design as Displayed at the 'Britain Can Make It' Exhibition*, HMSO, London, 1946, pp. 93-95

aluminium scrap. At 'Britain Can Make It', it was perhaps the most enduringly satisfactory of the innovative exhibits, and even six years later, amongst the flood of new European and American design following the war, it took a Gold Medal at the famous Milan Triennale in 1951. This was only to be the first of innumerable prizes won over the next thirty years, as Race Furniture became one of the leading innovators worldwide in application of industrial materials and techniques to high quality furniture design.

One important contribution to effecting that 'revolution' of which Russell dreamed was a practical and technical little book by the designer Gordon Logie called *Furniture from Machines*, published in 1947, which discussed leading new ideas from the pre-war years abroad, as well as the best examples of new approaches at home. Advantages of aluminium and magnesium alloys, he observed, were their lightness and 'the ease with which they can be fabricated'.

> 'They are easier to stamp, press and draw than steel is, and they have the advantage over most other metals in that they can be worked with ordinary woodworking machines ... by changing the cutters or sawblade.' [12]

In the long term, cost advantages were uncertain. Meanwhile, however, 'abnormal conditions may give the light alloys their chance of establishing themselves in the furniture trade'.

Race's design, of course, had already proved the truth of that. Logie illustrated it with this description:

> 'Chairs with light alloy legs cast in a tapered tee section. [Logie fails to mention that the question of aluminium's potentially unpleasant surface was overcome by stove-enamelling it – CC] The chair seat is of rubberised hair on a cast light alloy frame which bolts to the legs. The fabric covering is removable. The chair backs have key-hole slots and are sprung into position on studs.' [13]

Solid sections in steel, as Logie noted, become 'unduly heavy', but 'where the light alloys are used in solid sections, completely new possibilities of design are opened up ... [and] make available any shape of section which the design may require.' [14]

It is interesting to ask, as we pass on to a design of exactly the same year from the United States, how far were those 'completely new possibilities of design' really 'opened up' here?

For all this chair's great novelty in comparison to its stolid wartime precursor, or to the Walter Raleigh-Chippendale items of pre-War, this chair is still generically very close to its timber predecessors. We still have here four straight legs, two of which rise above the seat to carry a back support. They are fixed to a ring, carrying the seat, in a relatively traditional manner. We still have 'body-contact' parts – the seat and back – whose comfort derives from soft-upholstery elements built up in the traditional manner with cloth and padding over a plywood form, and they are slotted or bolted onto the seat ring and leg structure much as before. Even the 'new slender legs', when it comes to it, are not that thin. Why this relative traditionalism?

This chair is an example of development within a strong traditional type. The change of structural material has had some dramatic effects on its form, as we have seen. In the end, though, aluminium has some close affinities

12 Gordon Logie, *Furniture from Machines*, Allen & Unwin, London. 1947, p. 129

13 Logie, *Furniture from Machines*, p. 131

14 Logie, *Furniture from Machines*, p. 130

with the timber it is replacing. Logie's comment about how it could be machined with timber working tools is a little hint here. After all, the same is not true of steel. Steel is far harder than aluminium. We have already referred to steel as 'springy', meaning that well tempered steel can be made extremely stiff and will return quickly to its original form after bending. Think of a good quality kitchen knife, for example. If that thin blade was made of aluminium it would simply bend if you tried to spread hard butter with it.

So aluminium is in many ways actually quite close to timber in its properties under stress. Its introduction produced a dramatic change in form, but it was not so dramatic as the effect which steel would produce in its new post-war development. And there was another difference, in the intentions of the designers and the expectations of their clients.

C: RADICAL EXPECTATIONS: LCM SIDE CHAIR, CHARLES EAMES, 1945

In the forty-five years since this chair was designed, we have become so accustomed to legs and moulded plywood as thin as this that it is almost impossible to perceive what a radical breakthrough this chair was in its time.

FIGURE 34
CHARLES EAMES, LCM SIDE CHAIR, 1945

It was designed, as you see, in the same year as the Race BA chair, but not put into production for a couple of years. That was partly because the production techniques needed perfecting to execute such a radical structure; partly because there was not the same urgency in the USA as in Britain for rapidly restarting industry, refilling an empty market-place and in Race's case, seizing the moment of a technical and price advantage with that once-off supply of scrap material. But this delay like the differences in the two chairs was mainly because Eames was a very different kind of designer.

Where Race before the war was an unpretentious junior designer in a large industrial firm, the slightly older Eames was already a well known name as an independent design consultant and as a professor in one of America's most experimental schools of art, architecture and design, Cranbrook Academy.

Not itself directly touched by the war, American industry and its domestic markets were in a more normal state than the British and European in 1945. Eames resumed his consultancy, but was not himself becoming a manufacturer. He had to find a manufacturer. Race was in a position close to that we saw earlier, of the designer who also makes the product – in this case not with hand tools, but with a small factory specially set up to produce his design. Eames was the design consultant, who had to find a client to perform the manufacturing and selling operations for him, and in this case the eventual client was a well known quality furniture company called Herman Miller International.

The ultimate effect of these differences on the design itself, however, is our central interest here. The intention in this case was not to get something workable out quickly, but to produce a 'designer's item', something self-consciously innovative which attracts attention in a sophisticated sector of the market. The fundamental innovations were technical: the creation of a product with new practical merits through applying materials hitherto not used in this field. But with that went the desire to communicate the newness and the radical technical achievement aesthetically.

The four legs here are no longer 'sticks' attached to the edge of a seat unit. They are a self-contained structure using the minimum possible amount of a high performance material. As far as possible, the aspiration is to make the seat and back seem to float as if unsupported. To this end, as you can see in the smaller picture, the back is attached only by one single frame member, not the usual two. To express this floating quality even more strongly, the back legs do not connect directly with the seat either: they are a self-contained arch unit which is linked upwards only by the single slender piece of frame that carries the back.

The frame material is high-tensile narrow steel tube, formed with the minimum number of bends and then chromium plated. Chromium plating of course reflects the light and colours around it, and is famous for causing an object to 'dissolve' into its environment as a result. The back and seat elements are of specially moulded plywood. The double-curved form of each part achieves exceptional stiffness from very thin material. At the same time, its slight inherent flexibility, and the movement available to the sitter through flexing of the back on its single support, in particular, provide comfort by a far simpler means than the old fashioned upholstery used by Race. This is in fact not a dining chair, but as the picture shows, a lower, occasional chair. As such it is designed to achieve a level of comfort comparable to those heavily padded chairs traditional to entrance halls and waiting areas, as well as providing occasional extra seating for relaxation in a home.

In its way, this chair was as much a child of the war as the Race aluminium design. First, the new steels capable of this loading at this slenderness, and capable of being bent to such angles, were a product of wartime technology. Second, the seat and back elements exploited a new technology in which Eames himself had been involved during the war, developing a new type of leg splint for seriously wounded sailors in the US Navy. Traditional metal splints did not provide adequate support for a seriously wounded limb. At best they needed expert setting and at worst they caused gangrene. During 1942, Eames had worked on the design of special half-tubular elements, moulded to the forms of thigh, knee and calf in various sizes out of thin timber laminates, into which the leg could be laid for all round support. Perforated for ventilation and of quite complex shape, these were actually a greater technical advance than the chair seats in which he later exploited the technology during peacetime.

Many chair designs followed Eames' first venture into very narrow gauge chromed tube and laminated plywood, but in the determination to express visually what the technical achievement was, this remained the purest solution. It was a genuine rethinking of the elements of a chair where its British contemporary was only a modification of a traditional object.

Shortly we shall see another example of the same transatlantic difference, from a few years later. But first we look at another stage in Eames' play between the traditional and the new, and another means through which new design emerges, namely the design competition.

FIGURE 35
CHARLES EAMES, DAR ARMCHAIR, 1949

D: THE STIMULUS OF COMPETITIONS: DAR ARMCHAIR, CHARLES EAMES, 1949

Once the material and technology for making a minimal steel leg structure like this had been perfected, it could plainly be made to carry the person in a variety of different seats and backs. No longer were legs and seats constrained by each other, they could be solved as separate problems, and according to separate priorities, aesthetic as well as technical.

Eames' DAR chair was an example of an aesthetic idea preceding existence of the appropriate technology, of form being conceived before there was a material in which to make it. In my personal estimation it is a rather unsatisfactory object: unpleasant to handle, quickly becoming unattractive in appearance, and discordant aesthetically. To me, the two parts scream of precisely this different parentage. But it is a very famous item, generally considered a classic, and enormously important as the first use of another wartime invention, glass-reinforced plastic (GRP) to any consumer product.

The development of this design was also an example of the role of design competitions, though that process is not, of itself, a factor in the merits or demerits of the design.

As a major American propagandist for 'good design', before and after the war, the Museum of Modern Art (MOMA) in New York regularly organised open design competitions to help bring new ideas and new designers into the marketplace. As in any such competition, a topic would be set, sometimes functionally precise, sometimes more conceptual. Winning ideas and others might attract manufacturers interested in producing them, or might simply be published as provocative and challenging ideas. Both situations arose in the history of this design.

In 1940, before America entered the war, MOMA had organised a competition called 'Organic Design'. 'Organic' is a word of many meanings in design circles, sometimes referring to natural materials, sometimes to shapes and forms that are reminiscent of, or have properties of, forms occurring in the natural world. Eames and a colleague won first prize, with a concept for a chair that was a body-shaped minimal 'shell' on legs. It was not clear what material might be used to mass produce such a shell, but the image was seductive.

During the war period, extensive work was done in the States on the development of plastics for use in aeroplane construction, whereas in Britain the effort was mainly put into the use of lightweight alloys and laminated timber structures for this purpose. The material called GRP is a synthetic resin, inherently a brittle substance, reinforced with fabrics or 'mats' of glass fibre. The principle is the same as that whereby brittle concrete is reinforced with steel rods. By building up the form of resin-impregnated glass 'fabric' on moulds, complex shapes could be achieved of a strength not obtainable from plastics until this compatible form of reinforcement had been devised.

After the war, in 1948, MOMA announced another, more specific competition: an 'International Competition for Low-Cost Furniture Design'. Working with members of the Engineering Department at the University of California in Los Angeles, Eames now submitted a design which achieved the 'organic' form by means of the new plastic material.

It won a second prize, and his previous clients, Herman Miller International, took the design into production during the next year. When MOMA had an exhibition of Low Cost Furniture in 1950, the production model was on show, and the catalogue eulogised about it thus:

> 'This moulded fibre glass chair is in many respects an astonishing fulfilment of the ideas developed by Charles Eames and his occasional associate Eero Saarinen in 1940 when similar designs of theirs won first prize in the Museum of Modern Art's *Organic Design* Competition.
>
> 'Now it has been possible to find a plastic substance and a moulding process which allows this kind of shape to be produced economically. Chairs with complicated curves like this have always presented a special problem in the attachment of legs, but this time the problem has been solved with unusual directness and neatness. [Pads on the steel frame were screwed onto plugs cast into a thickened area of the plastic seat-bottom – CC]
>
> 'Perhaps the greatest advantage of this chair is the extraordinary lustre and soft, smooth surface of the plastic, which, strengthened by the silky threads of glass embedded within it, quickly absorbs room temperatures. Never before used in furniture, this airplane plastic is virtually indestructible and withstands stains and marks.' [15]

One failure of the design, in my view, is that this highly three-dimensional form has a very unattractive back side, as a direct result of the material and its production process. GRP is strong precisely because it is not just a plastic cast in a mould like molten aluminium, for example, but is 'laid up' on a mould as a mixture of glass and resin. As a result, only the side lying on the mould is smooth. The other side definitely does not have that 'extraordinary lustre and soft, smooth surface'; here the 'silky threads of glass' make it rough and uneven.

Thus the object which tries to be so sculptural and three-dimensional is not in fact three-dimensional at all. It has a smart front and a very unsmart back. Nor does the whole chair have that complementarity, the sense of each part contributing to a shared intention, which gives his LCM chair a coherence of form that in my view is more genuinely 'organic'. Vast numbers of these chairs came over to Britain, and you may have seen them. What was your judgement on their merits, I wonder?

15 Catalogue of 'Low-Cost Furniture' exhibition, MOMA, New York, 1950, quoted in: Carol Hogben, *Modern Chairs 1918-1940*, exhibition catalogue, Whitechapel Art Gallery, London, 1970, p. 27

E: CELEBRATING THE NEW ERA: ANTELOPE CHAIR, ERNEST RACE, 1950

Back in 1946, 'Britain Can Make It' had been a relatively limited exercise whose 'primary object', as Sir Stafford Cripps, President of the Board of Trade had said, was 'to prove that industrial design is by no means an impracticable and idealistic matter.' More than that, it was positive propaganda to further the process of public education which had been a by-product of the Utility scheme. Thus Cripps reasserted the then official view, that

> 'Good design can provide us in our homes and working-places with pleasant articles which combine good construction and fitness for their purpose with convenience in use and attractiveness in shape and colour.'

But as he also said, the exhibition's title showed 'a second purpose', directed at the country's potential export markets.

> 'After the trials and dislocations of the greatest war in the world's history ... [the] improvement of British design is an important factor – a very important factor – in our attempt to re-establish and increase our export markets.'

So much so, indeed, that to forestall inevitable consumer frustrations he warned the public 'Many of the goods that you will see in the Exhibition are not for the moment available in the shops', partly because 'they have not had time to get there', but mainly because 'it is more important for us that they should be used to purchase the food and raw materials without which we can neither live nor work'.[16] Thus export was having priority over home markets as part of the desperate effort of post-War reconstruction.

The Government was well aware that this situation, compounded with such measures as continued rationing of food, clothing and fuel, could make the prospect of peace seem even gloomier than war had been. Thus in the next year, 1947, it decided to celebrate the centenary of the Great Exhibition of 1851 with a nationwide festival of 'the Arts, Architecture, Science, Technology and Industrial Design'. Thus through a summer of performances, exhibitions and events, 'this country and the world could pause to review British contributions to world civilisation in the arts of peace.' 'Taken together', as the Guide to the eventual Festival of Britain put it, these events would 'add up to one united act of national reassessment, and one corporate reaffirmation of faith in the nation's future.'[17]

The centrepiece of this 1951 Festival of Britain was the South Bank Exhibition, where high technology and history mingled into an accessible and popular, but highly informative, entertainment. Nationwide, however, the Festival was the occasion not only for events and exhibitions, but a massive clearance operation to remove wartime debris; to clean, repair, repaint; to replant open spaces; overall, to return some colour and cheer to what had become depressingly drab environments in town and country. Near the Open University at Milton Keynes there are several villages which still bear the Festival of Britain emblem on their signboards, as reminders of how they won prizes in that exercise of restoring national morale. Maybe you know some which remain near you. It was a first official invitation to the nation consciously to relax a bit, and do some stock-taking. Outdoor

16 Sir Stafford Cripps, 'Foreword', in: Newman, ed, *Design '46*, p. 5

17 Ian Cox, *The South Bank Exhibition: A guide to the Story it Tells*, HMSO, 1951, p. 6

FIGURE 36
ABRAM GAMES, EMBLEM FOR THE 1951 FESTIVAL OF BRITAIN, WINNING PROPOSAL IN A LIMITED COMPETITION OF 12 DESIGNERS, 1948

cafés and parks opened up everywhere for the summer to give people a renewed taste of forgotten pleasures of ordinary community life – indeed pleasures which many had never tasted at all in the pre-War years.

The best younger designers were brought in to create the special displays, the special publications that went with them and the special style of the whole Festival. Its mood was lightness and cleanness, a sense that the clouds were lifting and the sun beginning to shine. It was popular, rather than elitist, but there was no conception then of anything that could be called ‘pop culture’. The two words were then seen as almost antithetical. Indeed, the Festival of Britain can be seen as one of the first steps, albeit unconscious at the time, towards asserting there was such a thing which was positive, not just a lowest common denominator. With the first Labour government newly elected, with new educational systems, the Health Service and so on being introduced, the underlying ethos was that high standards were now everyone’s right.

High standards of design were not peripheral to that. Today, when for better or ill such social programmes have been dismantled or revamped, and ‘design’ is again a word on everyone’s lips for spurious reasons of status marketing, it is only through these words of the time which I have quoted, from the wartime and after, that one can realise how genuinely and straightforwardly real design was seen as an expression of that effort of national reconstruction. Indeed it was seen as an essential contributor to actually achieving it.

FIGURE 37
ERNEST RACE, ANTELOPE CHAIR, 1950

All this was the ‘message’ conveyed by the Festival, and one of its most publicly visible manifestations, apart from the emblem, was the furniture used not only all over the South Bank site and throughout London, but for the exhibitions and the ‘festival cafes’ created on newly grassed street corners and bombed sites all over the country. The job went to Ernest Race, whom we have seen emerging earlier as someone technically inventive and able to capture the post-war lightness of mood. The result was this Antelope chair. The speed with which it became known colloquially as ‘the Festival chair’ reflected its success in capturing the Festival’s spirit and intentions, both practically and aesthetically.

The frame here was of steel rod, not the tube used by Eames. It was treated with a special rustproofing, then stove-enamelled in white, with little white plastic ferrules on the feet. Eames, you may have noticed, made these ferrules into a quite complex item, which tried to sit ‘flat’ on the floor and therefore had to be made to two special and different angles, for front and back legs. The solution here is simpler and cheaper: one standard ferrule for all feet, with domed base that contacts the ground at different parts of its surface on front and back legs. Quite apart from cost and simplicity, this rather generalised solution is far more appropriate for uneven outdoor surfaces than Eames’ refined precision.

The seats were of laminated plywood, of the grade called 'marine ply' whose resin glue withstands outdoor weather. The ply is a little thicker than in Eames' LCM chair, and is given a simpler form that curves up and down only in one direction. It was spray coated with an exterior grade cellulose paint in six different colours.

There is no attempt to conceal the fixing of seat to frame. On the contrary, those bolt heads in the seat are what give it visual life, marking the centre line of the pelvic support and the four outer corners with 'polka dots' that add to the overall delicacy and slight frivolity of the chair. You can see how well it goes with the emblem of the Festival, and the whole of its graphic design style, which was characterised, even in the lettering used and so on, by this same strong contrast of solid areas with thin lines.

If you compare the back to that of the Utility chair, you see the same absolutely simple means of support: straight upright bars at just the right angle. With the thinner material here, though, four of them are needed for even support across the sitter's back, not just three.

The bouncing curve of the simple side and arm piece allows a certain flexing of the back element, giving comfort by the means we discussed in the Eames LCM case. Although so light, it also creates a slight evocation of the traditional grandfather's wing chair, with those draught-proofing 'ear-pieces' as well as solid arms. Albeit subliminally, this made it feel a familiar object and a safe home-from-home, although to most people who would use the chair, its extreme lightness and openness were quite unprecedented.

Like Eames' LCM, the Antelope provided a seat which seemed to float, only here the impression was not so much that it floated in mid air, as that it was a rather maritime object. To British eyes, the combination of plain strong areas of colour with white stripes and bouncy curves gave it a nautical feel, a touch of the seaside air. The bunting in the Festival emblem evoked the same theme, and with bunting everywhere across the country that year, as the simplest way to make any street or kiosk seem festive, the chair was immediately at home in any environment. At the same time, the white frame against green grass evoked the elegant Edwardian garden furniture of the leisured, pre-war era.

Many thousands of these chairs were supplied for the Festival of Britain. The lasting merits of the design as a light chair, transparent but solid, for inside and out, led to continuing demand and production for over twenty years afterwards. Thus whilst embodying precisely the public mood and psychology of a given moment, it was acknowledged by public demand to be a classic solution to the general problem of the outdoor or lightweight chair in cheap modern materials.

F: MAKING IT ART: SIDE CHAIR AND DIAMOND CHAIR, HARRY BERTOIA, 1950 & 1951

We made a transatlantic comparison in 1945, now we shall make another five years later. It will show that the subtle differences of design intention and product which we observed before were more than accidental, that they continued to be characteristic of differences in design culture in the two very different post-war situations.

Race's Antelope chair, as we have seen, succeeded as much for arousing some profound historical associations as for its novelty. Indeed it was masterly precisely because it made such a clear and natural synthesis of these two dimensions of what the Festival of Britain itself was all about both technically and psychologically. In terms of the comparison we made between the legs of Race's earlier chair and Eames' LCM, the Antelope had caught up: these were not any longer 'sticks' stuck on to the edge of a seat, but were part of a self-sufficient filigree framework that 'received' a seat as a complementary item. But the seat was still a solid thing.

With the luxury of the more relaxed situation in the United States at that date, and in the spirit of gratuitous stimulation to invention for its own sake which we saw in the MOMA competitions of this period, the prestige furniture company Knoll International took a young sculptor, painter and architect called Harry Bertoia into their company in 1950, expressly to apply his artist's eye and imagination to a free exploration of possible forms for furniture. An almost exact contemporary of Race, Bertoia had trained in an Arts and Crafts School before studying under Eames at Cranbrook Academy.

Metal had been the main medium of Bertoia's sculpture, and as he worked towards the new task his first 'designs' were more sculpture than chair. The commercial value of something 'new' and 'art' in the US market was of course one of Knoll's incentives, but the market for sitting on bits of artwork was small even there! Then he came up with a chair which genuinely moved the current trend in furniture one stage forward, where 'legs' and 'seat' became even less distinct, both being made of the same material, fabricated in a similar way and only slightly differing in thickness. The first example was the Knoll/Bertoia side chair, soon known simply as 'the Bertoia chair'. The second was a development of this whose function was that of the 'occasional chair' we saw in Eames LCM, but with 'arms'. Logically enough, when you look at it, this was called the Diamond chair. In both cases the main frame elements were of chromium-plated steel rod and the seating shell was a welded and chromed lattice of thinner steel rods. Thin cushions, no more than shaped pads, were usually used on the sitting surface, but in principle were not considered essential.

Bertoia described the thinking which led to them thus:

> 'In the sculpture I am concerned primarily with space, form and the characteristics of metal. In the chairs many functional problems have to be satisfied first ... but when you get right down to it, the chairs are studies in space, form and metal too.
>
> 'If you look at them, you will find that they are mostly made of air, just like sculpture. Space passes right through them.'[18]

18 Bertoia, quoted in: Hogben, *Modern Chairs*, p. 29

FIGURE 38
HARRY BERTOIA, SIDE CHAIR, 1950

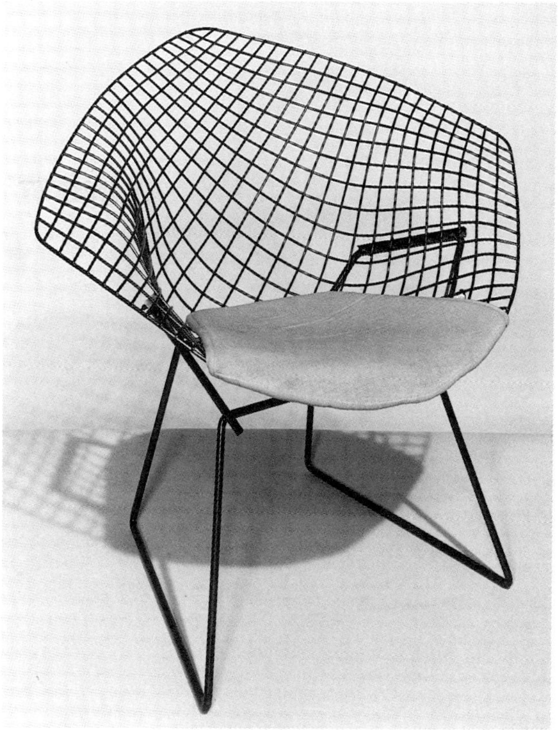

FIGURE 39
HARRY BERTOIA, DIAMOND ARMCHAIR, 1951

These chairs, like the Eames fibre-glass ones, also came to Britain in large numbers and you may have seen or used them. 'Classic' though they are considered to be in the history of chair design, my judgement on them is not so favourable. I'll tell you why.

Though they look light, they are actually very heavy, especially the Diamond. If you compare them with the Antelope, you will see that even the directly comparable Side chair has a lot of metal in it. The Diamond in particular is very difficult to move, except by just pushing it around. Across the 'arms' it is too wide to lift with ease, and if picked up from the back the weight makes it a most awkward thing. (The loose cushion also falls off immediately you tip it!) The Diamond is also very space consuming, but to no particular purpose, as the slope, low level and wide spacing of the supposed 'arms' make them virtually useless. The replacement of four separate legs by a kind of cat's-cradle also supposes a very even floor surface if the chair is not to rock on its base-bars. This design, in my view, really is more a piece of sculpture than a functioning chair.

As for the technical innovation, this too is more art than real. The fabrication is complex: many bends and spot welds. Like supermarket baskets or railway luggage trolleys, there is a nasty edge where the ends of the rods sit on the outer frame. And was this idea of total transparency really so new?

It is an interesting phenomenon of design, as it is for example of technology or pure science, that much the same idea can emerge in quite different places at about the same time. This is because it is a logical next step beyond the current state of ideas, and parallel developments can occur without either party being aware of the other.

Usually there are relatively few logical answers to a given problem, be it practical, technical or aesthetic, and different people may come up with the same proposal simultaneously but genuinely independently. Between this and direct copying there is a whole spectrum of connections. There is the process of subtle influence: something seen or heard that is not used immediately, but later stimulates a 'new' idea to emerge as if fresh from the subconscious. Then there is the direct and conscious influence, whether publicly acknowledged or not. And finally, there is straight copying or plagiarism.

The boundaries between these different processes are naturally a rich field for specialised industrial and patent, or copyright, lawyers. In looking at design, however, the most interesting aspect here is to observe how ideas do emerge either because they were 'in the air', or because they were 'blowing in the wind', through dissemination by such media as international trade exhibitions and professional publications.

Thus Bertoia's introduction of mesh was hailed as brilliantly new, but was it so? I suspect not. Widely featured in the British design press, and shown indeed at 'Britain Can Make It', had been a range of garden and conservatory furniture called Loggia, designed by a relatively unknown designer called Greenwood for what was then Britain's most international, design-conscious and innovative furniture shop, Heals. In 1947, Logie, whom we have already quoted, described it as

> 'An interesting new use of a traditional material. The frames are built up from wrought iron rods and the seats and backs are of wire mesh. The finish is sprayed cellulose enamel.'[19]

The range comprised a dining chair, a low armchair, and a splendid canopied sun-bed on wheels. The steel mesh of seats and backs was a little finer in gauge than Bertoia's and was therefore 'woven' rather than simply welded up in two perpendicular layers. In the British design of 1946, there was a lack of technical sophistication, and a lack of refinement in the forms, particularly in the curves, that Bertoia, five years later and across the Atlantic, had improved upon. Greenwood's chairs probably had the edge in comfort, however, as Logie could write that they were 'very comfortable, and the angles of chair seat and arm rest are particularly good.' But to most eyes his would always be an 'economy' type of outdoor furniture, whereas Bertoia had brought the mesh indoors, chrome-plated it and refined the form with his artist's eye to produce 'high design'. Indeed these chairs became international cult objects in their day, whilst Greenwood's remained provincial obscurities.

Who then had the key idea, and whose contribution made a 'design' out of a mere 'idea'? Such is the dilemma, as ideas and forms move around the world so freely that one is rarely looking at total originality from a single hand.

But to return to our main theme here: the Festival of Britain organisers sought a chair which was light in spirit as well as materials, new but also reassuring as part of a psychological effort amidst a mass public, and they wanted it from a relatively war-torn industry. In the same year, across the Atlantic, Knoll wanted a sophisticated, customer-catching item for an affluent and 'design' conscious section of the American market. Here, far from reassuring, the sculptor's eye must rework the form of an established item till it shocked just a little – and a powerful industry was waiting to execute his idea to the highest technical standards.

Thus at the same time, different cultural and industrial contexts, and different briefs from very different clients, moved the common pursuit of 'a lightweight steel-rod chair' in quite different directions on the two sides of the Atlantic.

19 Logie, *Furniture from Machines*, p. 117

16 DIFFERENT FORMS FROM THE SAME MATERIAL BY EXPLOITING ITS VARIOUS PROPERTIES

After looking at a number of designs in such a broad context, we shall now focus on one single aspect of design in isolation: materials. We have often mentioned materials before, but here I want to make one point only.

Any material has a number of different ways of behaving, all of which are at the disposal of the designer to exploit for his or her purposes. Sometimes, of course, the problem may be a negative one: of suppressing or inhibiting certain tendencies that will impede the functioning of the object. Either way, no material is so simple (or so dull!) that it can only be used in one kind of way, and hence will only logically lead to one set of forms. The point we shall observe here is precisely this: that when a given material is used in different ways, to exploit its different properties and potentials, the logical result will be different kinds of shapes in each case. We shall compare some chairs which give simple examples of this with four different materials.

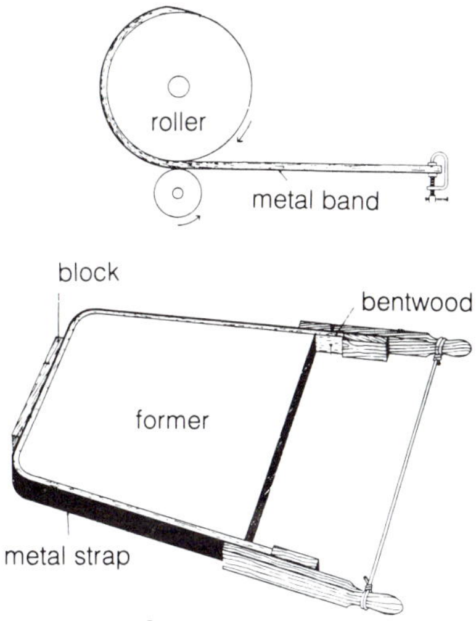

FIGURE 40
DIAGRAMS SHOWING USE OF A FORMER AND ROLLERS IN FABRICATION OF BENTWOOD CHAIR COMPONENTS. THE METAL BANDS AND STRAPS FORM A 'MOULD' FOR THE STEAMED TIMBER DURING DRYING

A: BENTWOOD

Wood is inherently an elastic material, which springs back to its original shape after bending. It has been well understood for many centuries that these bends can be made permanent if wood is softened in steam, then clamped around a form as it cools. This technique was widely used for making the frames of hulls and other equipment in the old sailing ships, for example. It was also used to produce furniture: the traditional English high-backed Windsor Chair used this technique. The British furniture industry became centred on the High Wycombe area precisely because of natural beech forests there which provided the ideal timber for this bending technique.

In the 1840s a furniture maker in Bohemia called Michael Thonet applied the technique to a new mass-market demand for lightweight mass produced furniture, in particular for upright chairs in cafes, restaurants, meeting halls etc. From his vast factories in Vienna, these chairs became familiar all over the world. Comfort came from the correctness of seat height and firm (though rather high) back support; from the slight depression in the thin, specially pressed, plywood seat element, or from the woven cane seat of more expensive models like that shown here.

FIGURE 41
MICHAEL THONET, CAFE CHAIR No.14, 1859

The frame itself is rigid. The legs are basically vertical struts loaded straight down the grain, where the wood is strongest. (Note where the tapering and thickening come, in relation to the earlier discussion of the Race aluminium chair of 1945, in Section 15B.) The bent forms are closed so that these too have virtually no flexibility. Even by today's standards, this chair is very economical of timber. When first introduced its lightness was revolutionary.

When Thonet and others sought to produce 'comfortable' chairs in bentwood, they did so by introducing upholstery or, in a famous Thonet design still much sought-after, by making a rocking chair. In all cases the actual frame remained stiff. However, in 1927, an interesting experiment was made in Russia to produce what was called 'a flexible Viennese chair'. The philosophy of the famous experimental designer Vladimir Tatlin was based on deeper exploration of the inherent properties and potentials of materials. One of his Moscow students, N. Rogozhin, developed a chair design which exploited the residual flexibility of bentwood as the source of springiness in a chair. Notice here first, that the timber is not thickened at any point to prevent bending: where it

must be stiff in the vertical 'pedestal' portion, a second member is introduced. Secondly, the curves are not closed in one plane to become stiff, they are open three-dimensional curves retaining the possibility for movement, so the principle used for support is not the stiff, straight strut, but the springy, curved cantilever.

This chair design was published in magazines and became famous in the West just as a picture, long since separated from its original caption. It is an interesting example of how we become conditioned to associate a shape with a given material that it was always assumed to have been a design for a steel chair. At that time, in the late 1920s, the cantilever form in chrome-plated steel tube was being widely explored by a number of designers (the one claiming to be first is discussed in Section 6 below). Even so respected a design authority as the Pompidou Centre in Paris had a replica made for its museum collection in chromed steel ... because no-one knew any better. Bentwood, after all is a stiff material, isn't it? Yes, but even the bent form retains enough flexibility to give spring if the appropriate shape and structural principles are used. When steel was short and banned for furniture during the Second World War, American companies devised spring systems for beds made of bentwood, in that case using mainly hickory, the wood used for hockey sticks.

FIGURE 42
N. ROGOZHIN, 'FLEXIBLE VIENNESE (I. E. BENTWOOD) CHAIR', 1927

B: PLYWOOD AND LAMINATED WOOD

Plywood is a development of the technique of veneering, already well known in ancient Egypt, whereby a thin layer of a high-quality wood is stuck onto the surface of a cheap one to give it a better surface. In plywood, the whole thickness is made of veneers, with grains in alternate layers running at right angles to each other, and each stuck to the next with glue.

In a thin form, as used for example for some Thonet chair seats, it is quite flexible over a significant length. In thicker form and over relatively shorter lengths it is virtually rigid.

Due to the effect of the two-way grain, it does not split if put in tension across the grain, like natural wood. Size for size it is stiffer in bending, and does not expand and contract under changing humidity and temperature.

The invention derives from the late eighteenth century but it was the nineteenth century before it was much developed, with the first American patent being taken out in 1865. We have seen the impact of the Second World War on developing certain new materials; for plywood, it was the First World War which brought it from obscurity into the front line of then 'new' materials.

After its widespread wartime use in aeroplanes, temporary buildings, crates, etc., commercial manufacturers marketed plywood energetically.

During the twenties and thirties it became widely used in furniture making, both as a cheap substitute for solid wood in the mass market and by innovative designers exploring it in its own right.

Not surprisingly, given their large timber industries, Scandinavian countries were centres of such ideas. The most experimental designer in plywood was the Finnish architect Alvar Aalto. His approach to furniture was the same as his approach to buildings. It was deeply rooted in a democratic philosophy which made available to everyone objects that combined high practical quality with an elegant simplicity, and transmitted certain spiritual, even political, values thereby.

In furniture as in buildings, Aalto's approach was based on a standardisation of individual components that allowed great diversity in the range of overall objects. This early furniture we see here is his simplest, and is typical of his design approach. One ingenious and elegant technique of lamination produced a leg form that made in two standard sizes, and became the basis of a great variety of chairs and tables for different purposes.

This is how he explained the vision underlying this approach:

> 'It may be difficult to give the general public a picture of the fundamental nature of standardisation. It is however still more difficult to explain its exceptional cultural value, for it may even be regarded as forming a corner stone of formal and material culture. It is to be hoped that in time it will become generally and clearly understood in our country that material building, which forms by far the greatest visible part of the cultural work of all peoples, definitely requires spiritual leadership which controls matter for the benefit of man himself in matters both great and small. [...]
>
> 'In the same way as words give a language its form, so will the internal working capacity of our country and the picture it presents externally, depend on what qualities govern its structural cells. These qualities will be partly responsible, too, for the characteristic form into which the life of our small people will mould itself. They can even have a bearing in such matters as individual liberty and the preservation of individualism, growth or suffocation, they greatly affect both the material standard of living and the spiritual structure of the country.' [20]

This simplest Aalto chair, and the even simpler stools related to it, were designed in 1929. The seat is a flat circle of plywood, sometimes faced in polished birch, sometimes in a high quality coloured and polished linoleum, the edge finished in a birch veneer. The backs of the stacking version are of birch bentwood. The back of the non-stacking one is a piece of moulded ply. In both cases you will note how low the back element looks, compared for example to that on Thonet's chair. But compare it also to the Rogozhin chair – and the back-sling we saw at the beginning of this discussion, in Section 12B.

What do you see? As in both those cases, the support is actually placed precisely in the lumbar region, where it is needed. When combined with the relatively low, small and, very importantly, *flat*, seats, this produces a chair which puts all pressure on the pelvic region and none behind the

20 Alvar Aalto, 'Suomen Arkkitehtiliiton Standardisoimistyo', *Arkkitehti*, 1943, 5-6, pp. 41 & 70

knees. The result is an extraordinarily comfortable chair for all seated tasks. It is instructive to compare this with the Thonet chair. Not only is the typical Thonet seat significantly higher, at 48-49 cm where the Aalto is 44cm, but the slightly concave form of Thonet's pressed-plywood seats puts the lower spine at quite a different angle; for the average person it also lifts the knees slightly above the pelvis, and produces a pressure point where the seat rim cuts the middle of the thigh.

The Aalto chairs were revolutionary in their time for this scientific approach to seating posture. As I have already hinted, however, there is another highly original feature here too. This is the way a rigid right-angled bend is made in the top of the leg by lamination, to produce a very simple screw joint between legs and seat.

Lamination is a cross between bentwood and plywood. It differs from plywood in an important respect. Whereas in plywood, as we have seen, the thin slices or 'veneers' of wood are assembled so that their fibrous natural grain runs at right angles in alternate layers, laminations are made of somewhat thicker slices assembled so that all the grains run parallel. There can be so-called 'cold lamination', but usually the whole lot is steamed before bending and gluing. The round heads of tennis and squash racquet frames were all made like this, for example, in the days before metal or the subsequent injection moulding technique came in. As such racquets show, the result is an extraordinarily strong and rigid form which consumes very little timber.

This chair leg is particularly interesting because it is only laminated where this is necessary for making the bend. The end of the solid birch strut is slotted, thin laths are inserted, and the whole is steamed, bent and glued. Notice how the two outer laminations run further down the leg than the inner pair. This is to eliminate the points of potential cracking where the bend starts in those thin outer tongues which are only supported on one side. It is a perfect handling of the transition from 'bend' to 'lamination' in a bit of birchwood, and a very famous little detail in its own right amongst designers because it is so elegantly simple. The result is attractive but the origins are scientific.

The proper use of a lamination, as you can imagine from the common direction of the grain, is where the stress runs in one direction only, parallel to that grain. Plywood, on the other hand, is equally resistant to tensions in both directions, because grains also run in both directions. In this sense, it is almost like a woven cloth in its capacity to take stress and to bend, provided, like a tailor or a dressmaker, you cut the right shapes out of the sheet at the appropriate angle to warp and weft. The third chair in this 'ply and laminate' group shows us precisely that principle in use, to make a seat-and-back element that is stiff in some directions and in some parts, and flexible in others..

This chair, called 'Jason', was designed in 1950 by another Scandinavian – a Dane, Carl Jacobs, who was then working in Britain. The base, as you can see in Figure 44, comprises four straight, slightly tapered circular legs made into a rigid frame rather like a trestle, by crossbars at the top. The upper two, at front and back, carry the seat. All this frame part is solid beech.

The seat-and-back element is cut from beech plywood. You see the 'paper pattern' for the shape to be cut from the sheet ply. The result looks natural and obvious, but it took 27 prototypes before exactly the

(A)

(B)

(C)

FIGURE 43
(A) ELEMENTS OF THE SIMPLEST AALTO FURNITURE, ALVAR AALTO'S STANDARD LAMINATED BIRCHWOOD RIGHT-ANGLE BEND FOR CHAIR AND TABLE LEGS, AND THE BIRCH PLYWOOD FOR SEATS AND TOPS

(B) ALVAR AALTO, 1929, GENERAL PURPOSE CHAIR, ARTEK CATALOGUE NO. 65

(C) ALVAR AALTO, SOON AFTER 1929, GENERAL PURPOSE STACKING CHAIR, ARTEK CATALOGUE NO. 68

right shape was achieved to obtain a seating element of exactly the necessary angle and degree of springiness without splitting of the timber. There is no heating or steaming here: a pneumatic jig pulled the 'arms' of the flat shape round into the required form, and the joint was glued to prevent springing back. (Look how these 'arms' interlock like a jigsaw at the ends, under the flat part: it's a structural equivalent to sitting on your clasped hands.)

This chair, like the Finnish one, was extremely strong. One exhibition catalogue later recorded how 'the manufacturers used to demonstrate this by having one large man sit on it while another stood on the rim of the back.'[21] The year after it was designed, this chair was widely used at the Festival of Britain: its curves and very direct, logical use of materials certainly give it something of the same cheerful but straightforward 'mood' we observed earlier in Ernest Race's Festival Chair.

21 Hogben, *Modern Chairs*, p. 52

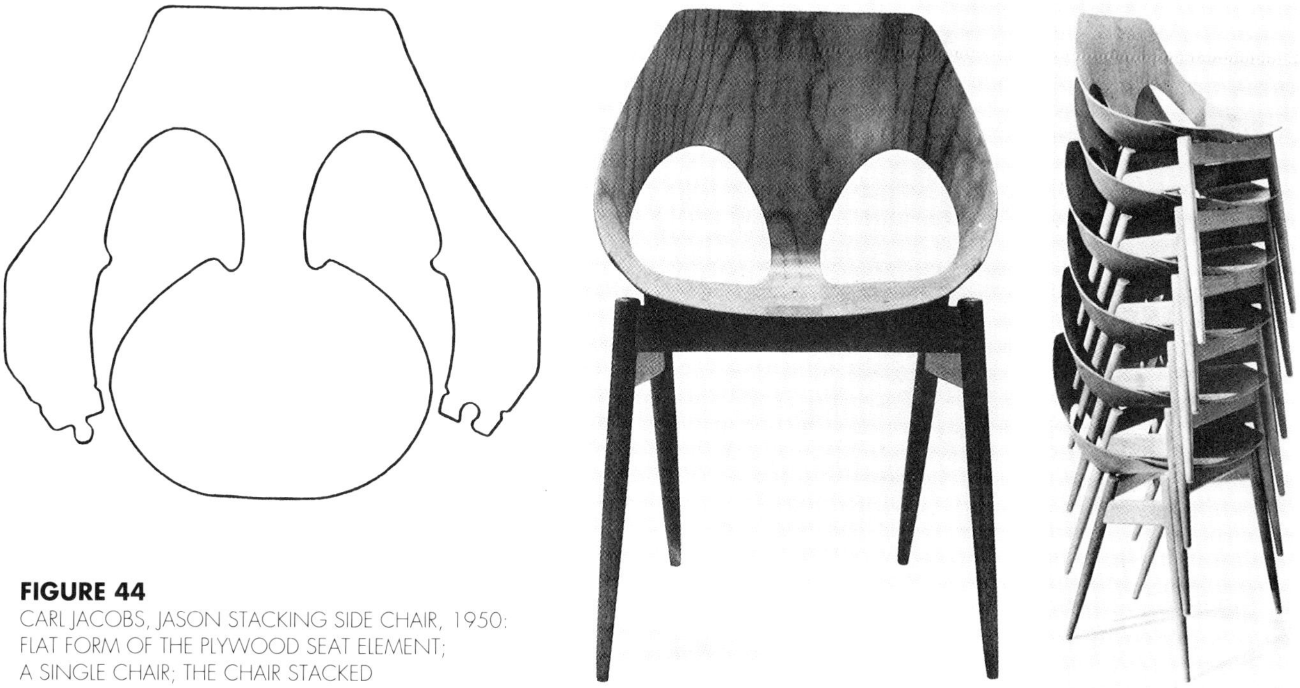

FIGURE 44
CARL JACOBS, JASON STACKING SIDE CHAIR, 1950: FLAT FORM OF THE PLYWOOD SEAT ELEMENT; A SINGLE CHAIR; THE CHAIR STACKED

In total volume, there is probably about the same amount of material in this chair by Jacobs as in the Aalto one. Before we leave this pair, look back at the photos of the individual chairs and try to feel the different ways in which wood is behaving in each case – how your own weight would be transmitted from seat to leg as you sat on each chair: where it would give comfort by *bending*, where it would give comfort simply by providing rigid *support* in exactly the right place.

Any designer who knows the stress and strain characteristics of different materials from scientific information, develops just such a 'feel' for how the materials will behave in different shapes and forms under different loading conditions. This is not some inborn 'intuition': it is the result of intelligent learning. But it is a very important aspect of the designer's expertise as a quick and successful problem solver. It is worth some conscious effort to develop such a 'feel' to a simple level for yourself, in the everyday things around you.

C: CHROMIUM-PLATED STEEL TUBE

Here is a material that is man-made – indeed it requires vast industrial processes for its manufacture. Yet at the level of basic principle, it behaves in essentially the same way as the naturally grown branches of timber we saw being used for bentwood.

Look at this pair of tubular steel chairs, and look back at the bentwood ones in Section 16A. You will notice immediately how this similarity of behaviour has led to similarity of forms. In each case, the first chair uses the material's *stiffness* in a certain shape, to carry the sitter's weight down to the floor through four straight legs at back and front. These are in direct compression: they do not bend. Then the second chair in each case uses the material's combination of stiffness with *springiness* in other shapes, to carry the load *either* at the front or the back, but not both. This is done through bending the material into a kind of 'S' to form a cantilever.

But there is one difference that your eyes will soon pick up. look how much simpler the steel versions are. No extra ring under the seat in the 'stiff' version (though look at the similar ways the seats are fixed to the legs, and the legs are curved over to make backs, in the two different materials). This absence of the lower ring reflects the fact that steel is much harder than wood: rigid joints can be made at one single point by bolting or welding, and they do not need the further support of a framework to hold the members in place.

In the cantilever form, it is not necessary to double the amount of material in the vertical part to prevent it snapping under bending with the sitter's weight. Steel is far less brittle than wood, and tube is even stiffer than solid rod. So a vertical 'leg' of similar outside diameter gives a nice little spring to the chair, and straightens up again when you get off it.

In discussing the Russian 'flexible bentwood' chair, I mentioned that this cantilever form in steel was fashionable amongst avant-garde European designers in the late 1920s. This example is in fact the first of them, initially conceived by the young Dutch architect Mart Stam in 1924.

It was characteristic of front-line young designers at that time, in the so-called 'avant-garde' of the twenties and early thirties, that the objects

FIGURE 45
MART STAM, CHAIR S.33, 1926

FIGURE 46
POUL KJAERHOLM, ARMCHAIR NO. 12, 1962

they designed derived from a distinct political and social philosophy which they hoped the objects, in their turn, would propagate more widely amongst the public. We have seen one example of this in the case of the young Finn, Alvar Aalto. Stam, his Dutch contemporary, exemplifies the European designers who sought to present industry not as a social enemy but as a positive and liberating element of everyday life. Light, dust-free, easy to clean and durable furniture like this brought the argument right into people's homes, offices and schools. It showed how new industrial materials could be an integral and beneficial part of their personal environment, just as industry itself was integral and beneficial to modern life as a whole.

For lack of technical resources, Stam's first models had to be made from straight lengths of tube joined by right-angle elbow pieces, thus losing the full springiness created by the continuous length of tube. Only in 1925 did he get access to the pipe-bending techniques necessary for making that component as he intended. Initially the seat and back were simple slings of canvas or woven webbing, but with the production version of 1926, which we see here, they became leather. Many other designers and firms have subsequently produced variants on the brilliantly simple basic idea. You have probably encountered some yourself, as Stam's innovation launched what became one of the established 'types' of modern chair.

By contrast with this, the form of the other chrome-and-leather chair here is very traditional in its concept, though designed nearly forty years later. This stiff tubular chair was produced in Denmark in 1962 by Poul Kjaerholm. It is worth turning back to compare it to Michael Thonet's stiff little bentwood cafe chair of one hundred years earlier: you can almost visualise one transforming itself into the other, with just a little borrowing from Comrade Rogozhin *en route*, for an idea about handling the back. What Kjaerholm does share with Stam is the use of leather to cover the seat and to wrap the top of the back, so the sitter's body does not touch the 'cold' or 'inhuman' structural material, steel.

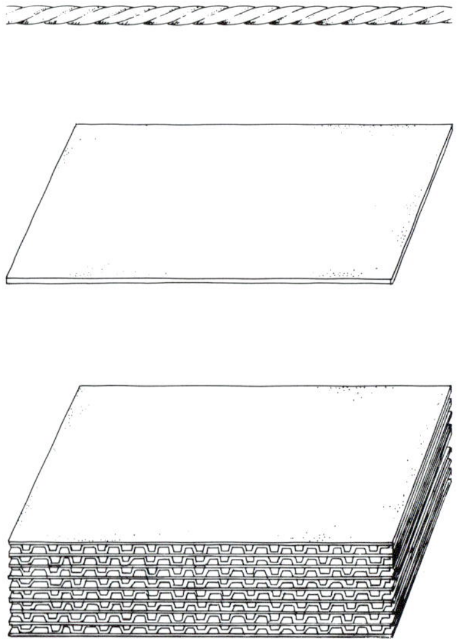

FIGURE 47
PAPER AS A MATERIAL: IN ONE-DIMENSIONAL FORM (STRING); IN TWO-DIMENSIONAL FORM (SHEET CARD); IN THREE-DIMENSIONAL FORM (MULTIPLE LAYERS OF CORRUGATED PAPER)

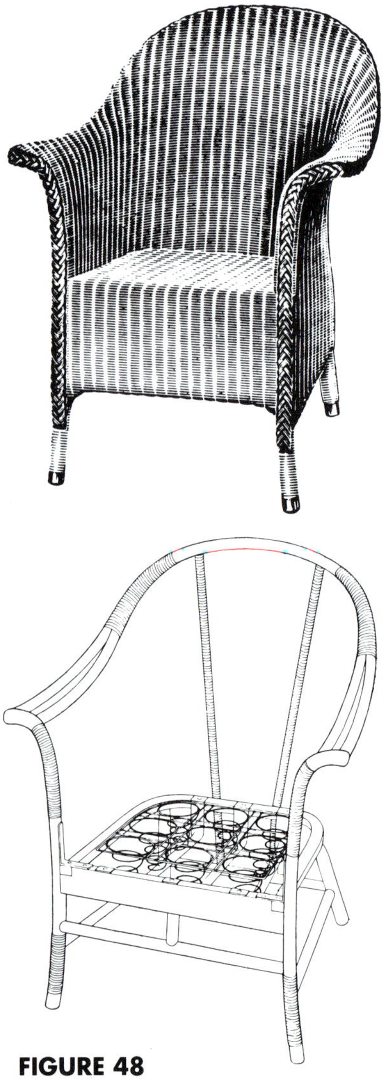

FIGURE 48
MARSHALL LLOYD, LLOYD LOOM ARMCHAIR, 1927: THE FINISHED VERSION COVERED IN WOVEN PAPER CLOTH; THE FRAME OF TIMBER AND BENTWOOD

D: PAPER

Yes, paper. Surprising as it may seem, paper has been used very successfully for chairs, and in three different forms: as a 1-dimensional material, a 2-dimensional material and a 3-dimensional material.

What does all that mean, for a start?

Something 1-dimensional is a just a line – a form with direction but no width. In this case, we mean a thread – a string made of twisted paper. Something 2-dimensional has direction and width but no thickness. It is flat, or just a plane. In the case of paper this means a sheet.

3-dimensional paper then sounds rather odd, but paper can be made into a 3-dimensional mass by corrugating. When lots of layers of corrugated paper are stuck together like the sheets of wood in plywood, shapes can be cut out of this thick 'board' with a bandsaw, just as they can out of a thick plywood.

This last one sounds as though it could eventually be strong enough to make a chair, but what about the ball of paper string and the flimsy paper sheet?

I suppose one could weave a hammock from paper string, and that is indeed a purer use of the material than the commercially developed one we have here. But few people realised this famous chair was actually paper.

Lloyd Loom chairs are a British product that has recently resumed production, thanks to a sudden upsurge in their second-hand value as their merits were again recognised. The technique was developed in the late 1920s by one Marshall Loom, from the old-established tradition of hand-woven cane furniture. In this paper-covered chair, the underlying structure is of bentwood. Sometimes, as in the example illustrated here, the seat had a full system of steel springing beneath it like an upholstered chair. The great feature of these chairs was their lightness, and they were initially developed for use in large numbers on the great passenger liners which provided mass international travel during the twenties and thirties. At the height of their general popularity in the later thirties, several factories were in production, and one alone is recorded as producing around 300,000 chairs in this material per year.

This is how the technique worked:

> 'Swedish *kraft* paper of very fine grain is used in rolls which are cut into thin slices rather like ordinary office gummed tape. These are unwound into a long strip which after passing through a gum bath is twisted through a die into a thread which is wound onto spools. This thread forms the weft of the fabric. The warp is similar except that it is reinforced with a core of 18 gauge steel wire.
>
> 'The fabric is woven in strips up to 40 inches wide. ... When covering a frame a piece of fabric is cut roughly to shape and moulded over the frame by hand, nailed all round the edges and trimmed off. Rough edges are covered by a braid of the same material ... and the finished chair is sprayed with enamel [paint, and dried] in a heated tunnel.' [22]

22 Logie, *Furniture from Machines*, p. 108-109

The result was firm but comfortable and very hardwearing. The same description comments that 'its one defect is the difficulty of cleaning in between the mesh, but this does not seem to matter in practice as the type of paint used does not attract dust and the texture of the fabric conceals it.' This was not a material for sterile environments, therefore, but it was brilliantly successful for public and domestic applications before the invention of plastics.

Paper 'in sheet' sounds even less promising than paper string for loadbearing. This famous example, also British, used a thin cardboard of special construction, and the chair was aimed mainly at children. But one principle is particularly well demonstrated by this object: that is the central importance of shape in determining how much load a given amount of material can support. The flat paper cut-out will plainly support almost nothing. Strength is created by the geometrical properties of the three-dimensional form into which it is bent or folded. If Carl Jacobs experimented with 27 different prototypes of the cut-out for his bent-plywood seat in 1950, this designer probably went through as many variants of his paper cut-out a decade and a half later, before curves, angles and assembly details were all just right for optimum strength and comfort.

FIGURE 49
PETER MURDOCH, SPOTTY CHILD'S CHAIR, 1964: THE CUT-OUT OF LAMINATED CARD BEFORE FOLDING; THE COMPLETED CHAIR

At one level, this was a 'fun' item typical of the 'Swinging Sixties': it was produced by the industrial designer Peter Murdoch in 1964. But appearances are deceptive. It was actually a much more serious object than that, both technically and socially.

Technically, the experiments may have started with a sheet of ordinary card, but the production model was sophisticated. As an exhibition catalogue of 1970 described it,

> 'This chair has a washable finish and makes a strong-wearing seat which will stand years of use. Its apparent simplicity conceals extremely complex development. Three different papers are used to make a board of five laminations.' [23]

Its social potential was the result of its extreme low cost, and the ease of transportation to any place in the world where many children needed seating:

> 'With a unit cost measured in pennies, a chair of this kind can be completed on one machine at the rate of one per second, and can be shipped in a minimal volume – 800 chairs in a pile no more than four feet in height.'

With all those advantages, what would you consider its main drawback? Most obviously perhaps the fact that it does not stack. But if space-saving is not a priority and cheapness or lightness is, this will beat the steel-frame-and-plywood seat hands down. Here is an excellent example of how no object can be judged as a 'good' or 'bad' design without reference to the precise problem it is trying to solve.

Low cost was also the motivation behind this third paper chair, produced for student use in 1972 by the Californian architect Frank Gehry.

23 Hogben, *Modern Chairs*, p. 69

FIGURE 50
FRANK GEHRY, EASY EDGES STACKING ARMCHAIR, 1972

Layers of corrugated cardboard were glued together, as described earlier, to produce a massive 'sheet' as thick as the width of the chair. With a large bandsaw on computer-controlled equipment, hundreds of these identical 'M' shaped 'slices' were cut off the block, and then spray-painted as the woven paper chairs were. The open 'M' form owes its strength as a seat to the several-inch cross-sectional thickness of the 'slice'.

That same thickness gives an air of domestic solidity to an object which actually contains almost no material at all. Like aerated chocolate there is lots of bulk with not many calories! The bulk also makes a solid space divider from an item which can be stacked up to the ceiling in perfect safety when the floorspace needs to be cleared.

What disadvantages might this item have? Rather combustible perhaps? An interesting question, but compare its likely behaviour to the speed of combustion and lethally poisonous gas from plastic foam upholstery.

In these examples we have looked at the shapes produced for chairs in response to the different properties of four different materials: bentwood, plywood, steel tube and paper. When we have considered function here we have meant practical categories of function, like dimensions to fit the human body, cleanability, lightness for transportation, strength under loading and durability in use. I mentioned the way leather was used 'to feel good to the touch' on the two chairs of tubular steel, but neither of those, any more than the rest we have seen, showed more than a passing concern for what we might call a 'sensual' function of the chair, far less for a real indulgence in sensuality. Suppose though, you wanted a chair to be 'sexy'. Suppose its function was only secondarily for taking the weight off the customer's legs, and was primarily for creating some sensual evocations? How then would you approach the task, and what would be the key variables in the design?

17 THE SEMANTICS OF DESIRABILITY

In earlier examples like the 'Japanese sitting machine' or before that, in such imaginary examples as the pop-group coloured cake, we have recognised that many objects have a function for the mind, a philosophical or psychological function, just as much as a function for the body, that is a physical function. The objects 'give off a message' as well as giving physical support, nutrition or whatever, to the body.

If anything gives off a message, if it communicates, this is necessarily done through some medium which makes it possible for our various human sense organs to pick up its signals. These senses are our basic equipment for taking in information about our environment: the classic 'five senses' are of course sight (eyes), hearing (ears) senses of taste (mouth), smell (nose) and touch (skin).

In each case we receive information in some coded form, in some 'language'. It is not an actual piece of leather which gets into our nose but the vapours it gives off which are picked up by our nose, which transmits signals to the brain that if correctly recognised, i.e. matched to previous experience, will tell us 'this is natural leather'. Likewise it is not a piece of steel that enters our eyes, but a pattern of reflected light rays, which again have no meaning until matched to prior experience by our brains. There is thus no such thing as 'pure seeing', 'pure smelling' and so on. It is always 'seeing and thinking', 'seeing and interpretation', 'seeing and recognition'. The two stages usually happen only microseconds apart, that is effectively simultaneously. But it is important to recognise that there are two processes here, corresponding to 'the object' and 'the person', for persons are very variable though an object may stay the same.

In trying to control and determine people's reactions to an object as potential customers and users, the designer has to grapple with this variability and complexity. At one level, every person's 'prior experience' is different. The designer is obviously operating at a more aggregated level: trying to evoke recognitions and associations that are common to a significant market group. When the market is potentially international, even worldwide, this becomes more complex again, thanks to differences in the 'meaning' of given shapes, colours, behaviour patterns, in different cultures, as a result of historically rooted cultural patterns. So, the wrong memory triggers may be awakened, that is to say, not those triggers which the designer intended. A form and colour combination which evokes say 'expensive exclusivity and dynamism' in Europe, for example, might give off the message 'cheap, common and rather flabby', in say, East Africa.

This whole field of visual psychology is very under-developed by comparison with the sciences which can predict the physical properties of an object – the load-bearing capacity of a given material in a given shape for example. At first sight it is the varieties of human experience which produce the greater difficulty here: every mind is shaped by a different bundle of experiences, whereas every 14 stone body puts the same 14 stone load on the chair legs. That is true, but in a mass market place in a given culture, conditioned by common exposure to today's mass media, there are substantial areas of uniformity. The greater difficulty is not the diversity but the subtlety: slight differences of curvature in a surface, for example, may cause it to evoke entirely different associations in everyone. This is where some of the 'art' aspects of design become important, in relation to mass consumer-oriented products as well as minority or more technical ones. The 'artist' in this sense is someone who has a finely tuned sensibility to nuances which are difficult or impossible to define objectively, i.e. to turn into a 'science'. And market share may depend crucially on getting precisely those nuances right.

Look for example at the subtly different shapes of such items as personal stereo (Walkman) equipment, or mobile tape-radio machines, aimed at the young market. Different curvatures of corners; different degrees of smoothness or protrusion of knobs; different directions and rhythms of movement involved in effecting control of the machine's output – sliding, twisting, pushing and clicking – and the different movements of the hand over the object's surface that result, quite apart from its textures and colours: all these are being manipulated by designers to produce 'desirability' in these items.

These kinds of equipment offer interesting examples, because they are bodily and psychologically very intimate accessories to the lifestyle of individuals in the market concerned. But the same kinds of factor operate in almost all other artefacts which are in any way an extension of our own bodily capacities – be they hand tools, power tools, or big machinery, for example. Furniture is part of that environment (and I need hardly mention cars).

One fact I have tried to stress, because it is often very inadequately recognised, is precisely this dual nature of 'function' in an object. As sophisticated social animals, human beings only get full satisfaction from an object which addresses mental or psychological requirements as well as purely bodily or physical ones. If you examine your multiple purchase decisions honestly, you cannot fail to recognise the truth of this. If you are on the other side of the fence, as a manufacturer (or the manufacturer's designer) concerned to sell in a market place, you ignore this at your economic peril.

There are two broad ways in which such associational factors work. We may recognise a whole form that 'reminds us' of some other form: there is 'recognition', literally reknowing of something already once known. Often the source of the association is quite consciously recognised. Or there is a more diffuse process of multiple stimulations and recognitions that may not be identifiable to the consumer individually or specifically, but which give signals to one or more of those five sense that add up, as a package, to association with some mood, environment, set of values or sensations.

We may call these 'literal' association and 'subtle' association. The general field of how symbols acquire and communicate 'meanings' is called 'semantics'. For a simple and I hope universally comprehensible example (within our cultural context) I have chosen two 'one-person seats with back support' which play with the 'semantics of desirability', i.e., which try to project themselves as 'sexy', the first by using 'literal' devices, the second by using 'subtle' ones.

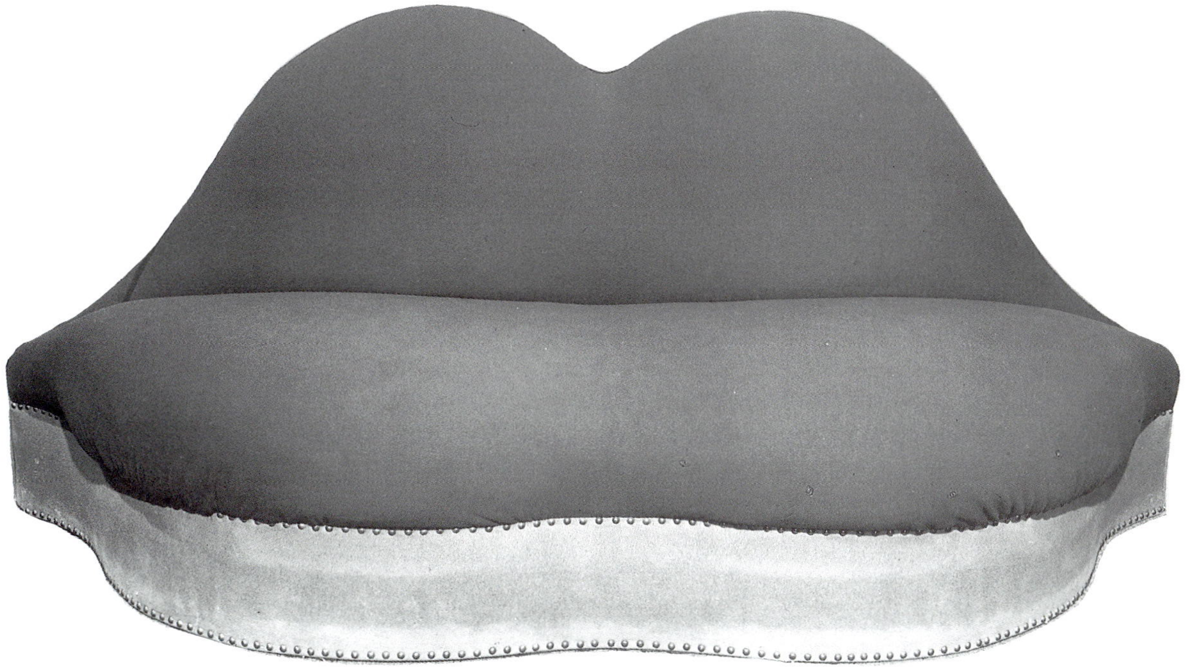

FIGURE 51
SAVADOR DALI, 'MAE WEST'S LIPS' WALL SEAT, C.1936

A: 'MAE WEST'S LIPS' WALL SEAT BY SALVADOR DALI, C.1936

Here we immediately see the limitations of the literal approach. Mae West's lips may have been the most desirable surface any man could dream to touch in the mid-thirties, but these particular curves probably evoke nothing if you belong to the generation whose sex-symbol is Madonna ... or Bardot, Monroe or Taylor. The colour of Dali's creation was equally specific. The satin upholstery fabric was in the lipstick colour of Paris couturier Schiaparelli called 'Shocking Pink', a brilliantly intense and slightly purpley dark pink that was the height of elite fashion at the time, precisely because Mae West always wore it. So colour and form went hand in hand here to create an instantly recognisable object – 'super-real' or Surreal as its designer called it, and almost a parody of the object which appeals by 'literal' associations.

B: 'MARILYN' CHAISE-LONGUE BY HANS HOLLEIN, 1985

Dali was of course a painter by profession, and this type of literal 'representation', in Surreal juxtaposition or scale, was his habitual medium of communication. Hans Hollein, by contrast, is an architect known for his sensual use of rich natural materials in the details and decoration of his buildings. His method works by subtle evocation rather than literal representation.

In this chaise-longue (Fig. 52) the basic line and proportions in themselves evoke a reclining female form, head above the rising back to the right, a hand languorously placed on a thigh to the left. Where Dali might have given us the hand itself, we have a form any hand would be irresistably drawn to caress: a superbly glistening sphere of natural hardwood. Look how the fleshy curves of the back squeeze another piece of curvaceousness to the right. That separate sausage of upholstery is in flesh pink: is it still the temptress's satin underwear or is this already her flesh itself? The rest of the upholstery is in a pale eau-de-nil, and both satins have a slight regular texture of squares in their sheen, giving the surface a pore-like, cellular life. Hairlike piping that delineates the edges of curves is half-way in colour between flesh and the deep golden glow of the burl wood base.

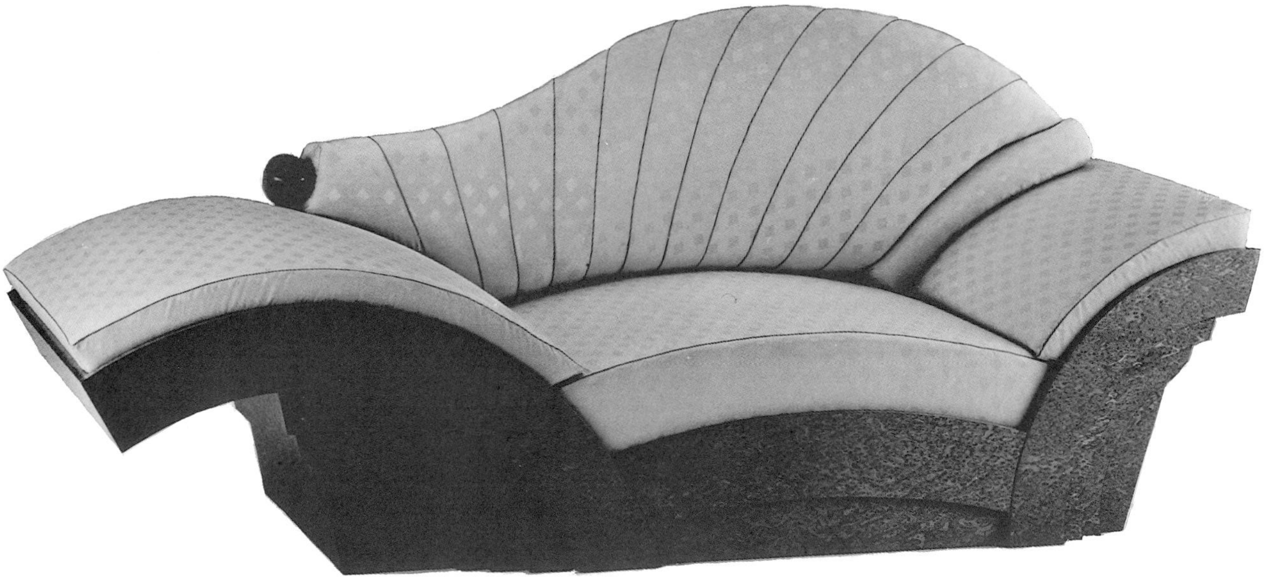

FIGURE 52
HANS HOLLEIN, 'MARILYN' CHAISE-LONGUE, 1985

This burl wood has an intricate grain which looks more like an overall texture than a conventional linear wood grain. With the curved ribs of its bold decorative layering it evokes the rough outside of a shell, opening to expose the contrast of its lustrous pearly inner lining. Deep in our memories this inside-outside contrast plays with evocations of intimacy, perhaps with images that have themselves built on this symbolism like Botticelli's 'Birth of Venus', one of Western culture's most abiding evocations of femininity.

Thus through colour, line and tactile qualities (that is, the quality of what we touch), Hollein's 'seat for one' plays with several of our senses simultaneously, whereas we rapidly tire of Dali's one-line joke.

Unfortunately we cannot show you these in colour, but if you have a spare minute try sketching them both for yourself with coloured pencils and you will soon understand even better the difference of subtlety.

PART 3

RECAP AND CONCLUSIONS

18 THE LESSONS OF IT ALL

All these chairs have been interesting, I hope, in their own right, as examples of different solutions to a broadly similar problem called, in the dictionary definition, 'a separate seat for one, with back'. The really interesting question however, for which I hope you have got some feel by now, is why they are different? Why is there not one 'right' chair for all manufacturers, all users, in all places and at all times? And by extension, one table, one kettle, one calculator, one vacuum cleaner, one car, one aeroplane and so on? Even when we have got those things nicely classified into 'types' like this, why are there so many varieties within the type – so many variations on the theme?

If you flick back through the Block, the pictures may be enough to remind you of the main reasons why each one of those chairs was like it was. Whose idea was it to produce 'a new chair'. Why did a particular designer, manufacturer, retailer or future user embark on such a venture at all? And which of those people was the initiator in each particular case, and with what motives?

Until we understand that, it is impossible really to understand what the design thought-processes, the logic, has been in each case. Whether the resultant object is 'good' or 'bad' has to be judged in relation to those intentions of the players in the game, as well as by certain more absolute criteria which they may or may not have considered, like whether it performs its intended tasks, is safe, is profligate or economical with communal resources, and so on.

Features in these latter categories are often obligatory elements of the thinking. For reasons that go back to social and political factors, all products must conform to them, within the degree of flexibility and choice left open by the criteria or regulations concerned. Standards laid down in this country by the British Standards Institution, the BSI, come into this category. You will notice products of many different kinds labelled 'Conforms to' or 'Produced in conformity with' a certain British Standard (BS). Other countries have their equivalents. DIN, for example, well known to photographers worldwide in relation to film speeds, and also in engineering, stands for the German equivalent of the BS: Deutsches Industrie Normen.

These standards for safety and other aspects of practical functioning are a vital aspect of industry's public responsibility. Designers must know the terms and implications of those constraints intimately, in order to work imaginatively, as well as responsibly, within them. But precisely because they apply to everyone in a given situation, they are not the first and primary source of the differences between things. The differences, as you will have seen by now I am sure, arise from different intentions, needs and situations of the people involved in the life-cycle of the object: they make it, they sell it or they use it – and in the middle someone brings all those people's different priorities and requirements together in designing the object that passes along that economic chain.

Before we started looking at all these chairs, I stressed that they were not chosen at random. I hope that some very specific points came over to you in the reading. In these last stages of this Block, and thereafter in your TMA work, we shall sift through that material again and again, to extract the maximum juice from it. These descriptions of chairs will form the 'encyclopedia of possibilities' for your own design – precisely in order to limit the territory and make a rich learning process (and I hope a fun activity!), from what could otherwise be an impossibly open-ended design task at this stage.

The first stage of this is to look at our examples in terms of this most fundamental determinant of 'why they are different'. That is, as I have said, in terms of the question: 'who was the initiator of the whole process, and what was their motivation?'

19 PEOPLE INVOLVED IN THE PROCESS, AND THEIR ROLES

Near the beginning of the Block we had a discussion called 'The designer's relationship to the producer and the customer'. We shall now return to that topic, as part of our analysis of the lessons of these chair histories adding a little more detail, but also abstracting the presentation of the various sets of relationships into systems diagrams, rather than the pictorial ones with which we started.

We saw that in the normal situation today there are essentially three key people or groups of people involved in the emergence of a new product into the marketplace: a maker, a designer and a user.

We also saw the situation which normally prevailed before mass production, when for items like furniture or agricultural equipment an individual user would go directly to an individual with suitable craft skills. They would discuss any slight variants required on the standard type, and the object would be made to the user's order. This user was related to the maker as customer or client, according to the extent to which consultation took place, or whether a mere order was placed for a standard object. (Look back at Part 1, Section 6 if you want to recap on my definitions of those terms.)

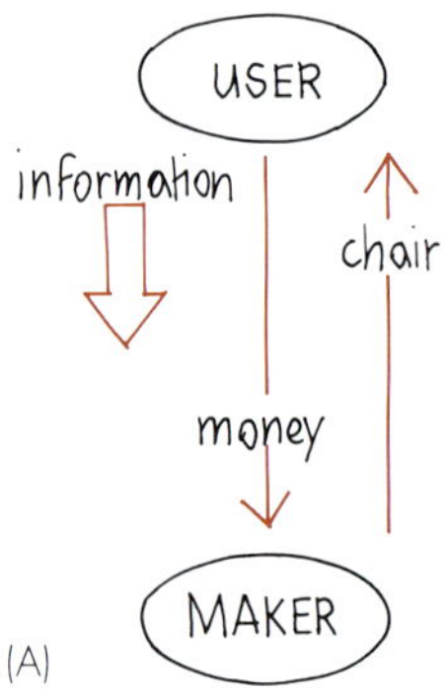

(A)

Within that situation, certain more innovative 'designer-craftsmen or craftswomen' emerged, to fulfil the desires of those affluent levels of society which could afford to give high priority to aesthetics and fashions. The man called Adam who devised the Coronation Chair in 1300 was a very early example of such a designer-craftsman. A little closer to mass-production were those famous names of the eighteenth and early nineteenth century like Thomas Sheraton, whose 'gentleman's reading chair' of 1802 we saw earlier. They generally made to order, but once their name and style was well known, items and sets would also be made 'on spec.', poised to be ready for the kinds of order that their fashionable status made fairly predictable.

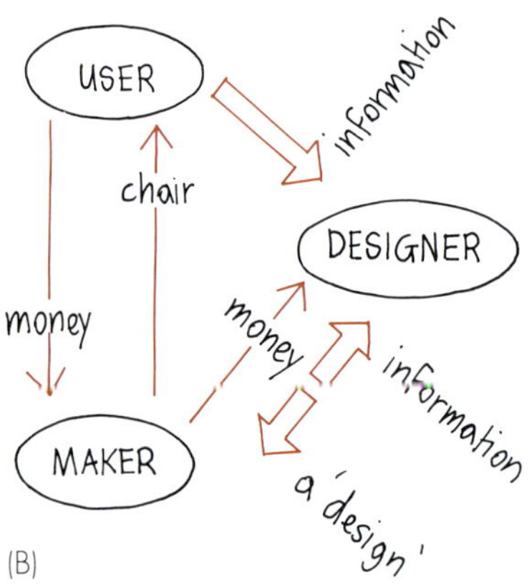

(B)

FIGURE 53
(A) PARTICIPANTS IN THE CRAFT PROCESS OF DESIGN AND MAKING (COMPARE THE VERSION WITH PICTOGRAMS IN FIGURE 10B)

(B) PARTICIPANTS IN A SEMI-CRAFT OR EARLY MASS-PRODUCTION PROCESS OF DESIGN AND MAKING (COMPARE FIGURE 10C)

By the time mass production techniques came in, and industrialising countries had a substantial population affluent enough to represent a market for the speculative manufacturer, there was increasing need, as we saw, for an intermediary in the producer-user relationship. There was a need for someone who could interpret the *likely* demands of *potential* customers, to whom manufacturers might hope their products would appeal over those of their business rivals in a marketplace. This was the 'industrial' or 'product' designer.

Whilst mass production was still relatively new, through much of the nineteenth century, it was quite normal for the manufacturing company also to sell the goods, through its own sales department and catalogue. The more diverse and numerous the products that became available, and the larger the potential customer base, the more essential for optimum operation of the market was distribution: producers wanted to reach the eye of the maximum number of interested customers; customers wanted to see all available products of a certain type in one place before making their purchase decision. Thus independent intermediary undertakings developed to deal with mass distribution. These are the retailers. (In many industries there are intermediary wholesalers as a supplementary part of the retailing function.) These people make nothing, but display and sell the finished goods of numerous manufacturers at sites located for customer convenience rather than production advantage.

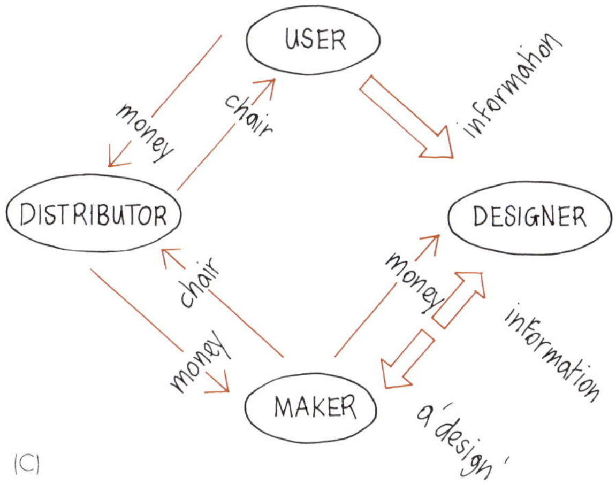

(C)

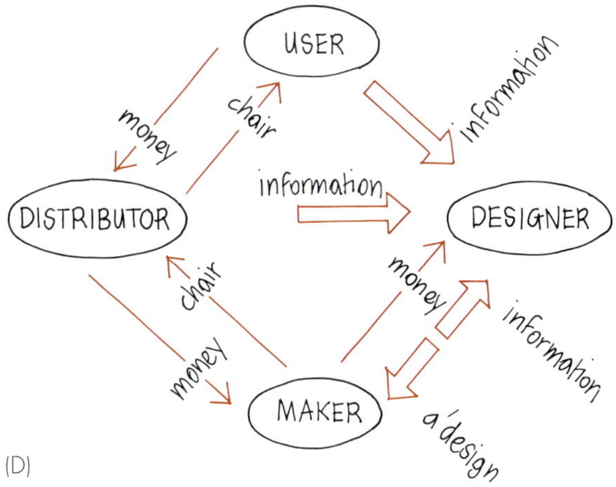

(D)

FIGURE 53

(C) PARTICIPANTS IN A MASS PRODUCTION PROCESS THAT INCLUDES A SEPARATE DISTRIBUTION FUNCTION (RETAILER PLUS WHOLESALER) FOR FINAL DELIVERY TO THE CUSTOMER IN THE MARKETPLACE

(D) THE SAME SET OF PARTICIPANTS BUT WITH FEEDBACK COMING TO THE DESIGNER FROM THE MARKETPLACE, VIA THE DISTRIBUTORS THROUGH SALES FIGURES, SHOP ASSISTANTS' REPORTS OF CUSTOMER REACTION, ETC. (COMPARE FIGURE 10E, WHICH COVERS THE CASE OF FEEDBACK COLLECTED AND TRANSMITTED BY A SEPARATE SPECIALIST MARKET RESEARCHER)

On certain occasions and in certain circumstances, distribution is not conducted through this free-market mechanism. Do you recall one such case from our chair examples? I am thinking of the Wartime 'Utility' chairs.

They were preferentially distributed on the basis of proven need as defined by the Government, to people bombed-out or setting up first homes in wartime. A parallel example today would be the distribution of medical products through the National Health Service. Payment is made by the user, and commercial retailers may serve as the Government's agents in the distribution process, but 'who has what' is determined by professionally agreed need, not simply by what an individual fancies and has money to pay for.

The primary function of the retailers is thus as distributors. This is the role they perform for potential users. But within the process as a whole, they have another vitally important role, as the source of feedback for designer and manufacturer about the market preferences of those potential customers. Such information on the nature of the objects that may appeal to customers, and the number of them they may hope to sell, is the key to their successful planning of future production. This information from sales figures and personal observations of retailers is complemented by broader based information from professional market researchers, who have contact with non-purchasers as well as those who have already had some contact with a retailer. The role and function of market research is discussed further in Block 2.

In the simple case represented here by those folding stools for the family caravan, a designer obtains the information necessary for conceiving the new product directly from the customer/client who enters the office, just as the craftsman did in the past. Amongst the categories of designer still operating in this way are architects, since larger buildings are still usually one-off designs for a particular location and client. However for the industrial designer who is aiming a product into the general retail market, those retailers, with whom customers conduct their decision-making and purchasing, represent one of the most important sources of information on how consumer preferences are moving.

It is not the one-off craft production situation which interests us here, of course, but the mass-production situation. Thus this question of information is central: indeed information is the very raw material of design. To design something, as we saw at the very beginning, means 'to make a mental plan' for it: in today's terminology it is a software activity, not a hardware one. When you start designing shortly in your TMA, you will experience this at first hand. Thinking it out – that is sifting information and making decisions – is nine-tenths of the job (and the difficult part). The physical 'making' of an object (in a little scale model, for example) is relatively easy when the designer has already made all the decisions. (You may like to test out this proposition when you come to make your own chair design in the TMA!)

20 MOTIVATIONS AMONGST THE CHAIRS: A REVIEW AND AN EXERCISE

In order to make money (or indeed to perform any socially useful function at all), the potential maker is seeking to identify a niche somewhere in the whole information mechanism we call the 'market'. That niche may be an unfulfilled need, either physical or psychological; or it may be a technical or economic niche that could be exploited profitably to produce an item which already exists to better standards or by more economic means.

It is not of any importance in principle who first cries 'Eureka!' and identifies that niche. The suggestion may come from users themselves; from information gatherers like shopkeepers and market researchers; from the designer's observations, or from a manufacturer. It does not matter, because at some stage in his or her thinking the designer in a sophisticated modern society has to consider the interests of everyone in the process. The starting point, and probably the order in which the multiple factors are initially considered, will differ according to where the initiative came from. The priorities assigned to various factors will differ in each case, and in some situations may be significantly influenced by those of the initiator who set the whole process in motion. Some of those priorities may be 'zero', but that too, of course, is a decision based on information.

If the designer has to work through *all* the basic aspects of an object, what are these? We can frame them in general terms as six questions.

- For what purposes does the object exist?
- For whose benefit does it exist?
- How do all those various persons interact with it, physically and mentally?
- What shape is it?
- What is it made of?
- How is it made?

You will notice that I have not numbered these six questions. That is for a good reason. After reading the last bit of text, perhaps you can already deduce what that reason is? It is firstly because they do not have any fixed *order of priority*, and secondly because there is no one standard or 'correct' *order in which they should be answered* during the process of designing an object.

All those questions *must be asked*; they have to be looked at individually, and in terms of how each one interacts with the rest. There are so many of them, and so many possible combinations and permutations of the satisfactory answers in each category, that there can almost never be a simple route through the problem. Whatever the starting point, either in external stimuli or the designer's own thinking, he or she will start by working logically and carefully over the ground again and again in all directions, exploring the territory and identifying the stumbling blocks as well as the key factors for the solution. This process is very largely one of self-education about the problem, its constraints and possibilities.

Design, as a task and as a process, involves analysing a lot of different aspects of a problem and formulating proposals for an object which solves them all simultaneously, or within an agreed order of importance, in a way that reflects the priorities of the people concerned. Some bits of this process or activity are science or technique and can be learned fairly mechanically. Other bits of it are art: they are a skill which is partly aptitude, and is partly developed through practice and experience.

There is one characteristic feature of design which is often very difficult for specialists from different fields to grasp intellectually, and to handle. This is the fact that it inherently involves producing a synthesis of objective knowledge with subjective judgement. Indeed almost every stage and step of it involves somehow bringing together factors from these two inherently incommensurable categories. Thus someone concerned with *engineering aspects only* of a product will choose a material on the basis of specific, quantified information from materials science. But the designer proper, who is responsible *for the product as a whole* has to juggle together simultaneously these objective, scientific factors and such factors as the psychological or symbolic connotations of the resulting form or of the material's colour, which are highly dependent on cultural conditioning in the target population, and basically subjective. This lack of overall right/wrong answers is the fascination of design to many people, but the irritation of it to others. When you start to play with the TMA, it is precisely these interactions which you will be learning to experience.

Out of this preliminary exploration of problems and possibilities, the designer will emerge with an initial proposition. Indeed, the emergence of 'a first workable idea' is probably the sign to stop the general exploration and to start looking in detail at each dimension of the problem, often drawing on specialist expertise as appropriate in testing out the proposition in order to refine it further.

With luck each move that the designer makes, or each iteration of a particular little cycle of experiment, produces improvement. Sometimes that improvement comes through the painstakingly logical worrying away at the problem. Sometimes it comes as a sudden insight when the designer is away from the drawing board, having temporarily abandoned the problem to do something else for a while after many failed attempts at a solution. On yet other occasions a bit of work seems to produce no useful result at all – but then proves to have generated some vital experience or information for successful resolution of a later stage in the process. These various patterns of conscious and unconscious thought in problem solving are of course not unique to design: they are characteristic of any complex problem solving activity. But the professional life of a designer is so dominated by this type of problem solving that techniques for handling it productively have to be something consciously understood and used as part of his or her professional skills. These topics are the specific subject of Block 3.

You may already be alarmed at how maddeningly (even terrifyingly!) open-ended it all sounds. And it often is. But the essential skill lies in that 'logical and careful' process of multi-dimensional 'self-education' and exploration of the problem; in moving from that to a proposition; in repeated retesting and further decision-making.

So, the initial impetus to the process will affect the starting point of this process of designing a new product, and will probably also have a significant influence on the way the priorities are weighted at each step as the task proceeds. So how do all the parts of it interrelate? In the rest of this Block we shall work towards a diagram that helps us to understand these relationships, and to ensure that no area of the problem gets forgotten during obsession with any other area.

THE EXERCISE

In order to start mapping out a diagram of how all the various aspects of the task interrelate, let us look back at those chairs. We will start by looking at them in terms of the different kinds of initial impulse which set the whole process going in different cases, and which therefore influenced the balance of priorities within which the designer was answering those six questions. We forget 'whose idea it was', since that of itself, as we have seen, rapidly ceases to have much significance. We are concerned with what the actual idea or motivation itself was. After reading the Block these should ring some bells with you, as will the names of the chairs listed on the right.

You will probably notice immediately that these two columns do not proceed in parallel. By this I mean that the chairs on the right are not in an order which corresponds to the list of 'motivations' on the left.

In this first stage of our exercise in sifting and analysing the Block material, take a few minutes, a pen and a ruler, and link up the chairs with what seem, from your memory, to be the relevant motivations. There are more chairs than motivations, so some of the motivations will turn out to be appropriate to more than one chair.

Leave out anything on which you are stuck at present.

Then try to do a little thumbnail sketch of each chair from memory. My examples for the first two will show the primitive kind of sketch I have in mind here: just something that captures the chair's basic shape and character.

When you have tried to do both these things, go back to the text for a recap on any chair on which you were stuck, either in making the sketch or in recalling what motivation lay behind it. Fill all the empty gaps you can now.

My own version is shown on page 109. In some cases, there could be differing interpretations of the motivation: if we have not agreed somewhere, ask yourself why. But if you are happy with your reading of that particular story, by all means stick to your guns!

Those particular sets of motivation and initial impulse are of course a relatively arbitrary list, derived from a few particular case-histories. But they cover a wide range of areas. Indeed, if we look at them more closely, they cover most of the territory embraced by those 'six questions' we had just now. Let's have a look.

Take the first question on my list: 'For what purpose does the object exist?' In the broadest sense, and according to the way I have been arguing, all of the motivations listed here – or indeed any others – could be said to be 'the purpose for which the thing exists'. Its purpose is 'to fulfil this', 'to exploit that', 'to provide something', 'to achieve something', to encapsulate or express something', or 'to make a certain kind of statement'. In human life, these are all legitimate purposes.

It becomes easier to sort them out or classify them usefully if we move on to the next question: 'For whose benefit does the object exist?' Why does this help? Because these motivations are of very different kinds. They reflect the interests and needs of quite different people within that cycle of the chair's conception, manufacture, distribution and use that we analysed in those earlier diagrams.

Initial impetus and main motivation

To fulfil a specific practical task not otherwise answered

The desire to exploit a glut of a certain raw material on the market

The challenge of exploiting the properties of a familiar material in a new way

To provide the widest possible fulfilment of a social need

To achieve new effects by applying new technologies not previously used in this product area

To encapsulate and express a public mood at a given moment or event

To express or symbolise a specific philosophy or view of the world

To make an artistic 'statement'

Name of chair

Race BA Aluminium chair 1945

De Graeff Body Tool 1975

Clinch Utility chair 1942

Sheraton Library chair 1802

Bertoia Diamond chair 1950

Aalto stacking chair 1930

Rogozhin Flexible Viennese chair 1927

Dali 'Mae West' seat c. 1936

Race Antelope chair 1950

Nada-chair back-sling 1980s

Eames LCM chair 1945

Coronation Chair 1300

Eames DAR chair 1949

Your thumb-nail sketch

FIGURE 54

Of course the primary interest of every player in the first three stages (conception, manufacture and distribution) is economic: few people conduct businesses except for a livelihood, i.e. to make money. We are looking here at the next level of sophistication. Assuming the underlying aim is to make some money, how do they think they can capture market share?

Who is the most obvious beneficiary of the first aim here, which is to solve 'a practical task not otherwise answered?' Plainly the beneficiary is the user who has the practical requirement. That is the simplest kind of 'function': practical function. But earlier we recognised that objects have a second kind of 'function' in the lives of their users: a psychological function.

Let us return to the list and see if we have an example of that here.

Plainly 'exploiting a glut of raw materials' is not directly related to fulfilling a psychological function for the end-user. (At least, back in 1945 it wasn't. So complex are these things though that now, in the resource-conscious 1990s, it could indeed give some users great psychological or philosophical satisfaction to know that their chairs were made from a material that was unwanted, therefore cheap, and crying out to be recycled. But here we will stick to the facts of the time.)

It seems to me that there are three motivations in my list which quite directly fulfil some kind of psychological function.

Which three would you put into that category?

I had in mind the last three. In 'encapsulating and expressing a public mood', the object was obviously fulfilling a psychological function amongst many people, in a 'public'. The object that 'expresses or symbolises a specific philosophy or view of the world' could be fulfilling either an individual psychological need, or a mass one, according to how widespread is the 'philosophy or view of the world' concerned. If you notice, we have here one example of each of these. As far as Britain is concerned, the Japanese-inspired 'body-tool' addresses a very minority view of the world, whereas the Coronation Chair addresses a view of the world which is not merely mainstream, but is so official as to be established in the constitution.

So we have here two examples each of chairs whose main purpose is to fulfil specific *practical* and *psychological* functions which are not previously satisfied by any other object.

What about the third 'psychological' aim in my trio, that of 'making an artistic "statement" '? That seems to me to differ very fundamentally, because it is more concerned with expressing the personal world-view of the designer, than with identifying and filling some requirement of the user. Neither of the chairs in this category (see Figure 54) is actually very good for sitting on. Indeed, one could describe them both, in my opinion, as basically ego-trips for their designers.

Before we leave the user-oriented motivations, there is one in the middle I have called 'providing the widest possible fulfilment of a social need'. If we look at the two chairs which arose from this kind of motivation, neither is fulfilling any practical (or psychological) function that is 'not otherwise answered'. In practical terms, both are very traditional upright chairs for general use, mainly for working or eating at a table. What distinguishes them, if anything, is their simplicity and plainness, their

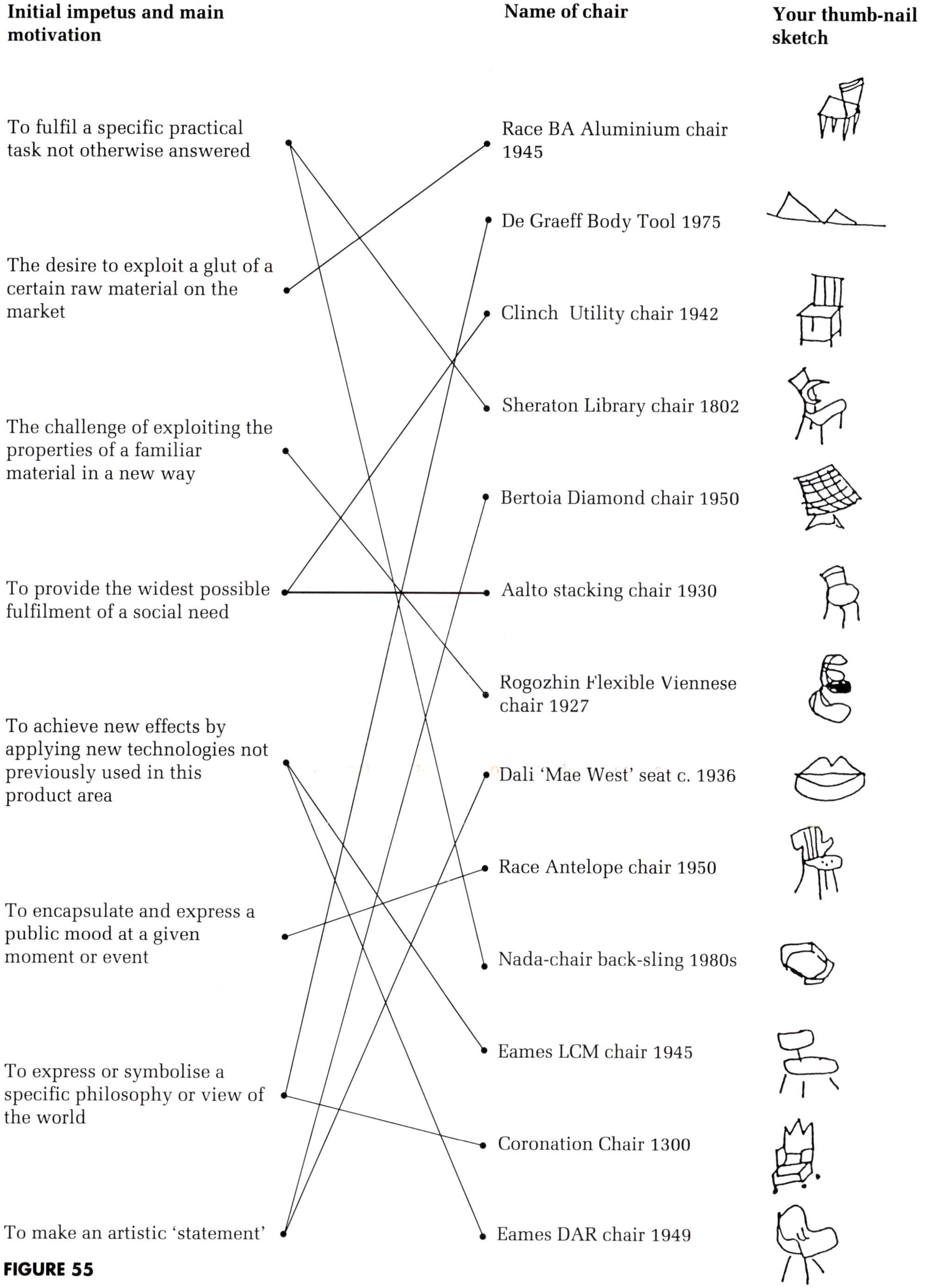
Initial impetus and main motivation
Name of chair
Your thumb-nail sketch
To fulfil a specific practical task not otherwise answered
The desire to exploit a glut of a certain raw material on the market
The challenge of exploiting the properties of a familiar material in a new way
To provide the widest possible fulfilment of a social need
To achieve new effects by applying new technologies not previously used in this product area
To encapsulate and express a public mood at a given moment or event
To express or symbolise a specific philosophy or view of the world
To make an artistic 'statement'
Race BA Aluminium chair 1945
De Graeff Body Tool 1975
Clinch Utility chair 1942
Sheraton Library chair 1802
Bertoia Diamond chair 1950
Aalto stacking chair 1930
Rogozhin Flexible Viennese chair 1927
Dali 'Mae West' seat c. 1936
Race Antelope chair 1950
Nada-chair back-sling 1980s
Eames LCM chair 1945
Coronation Chair 1300
Eames DAR chair 1949

FIGURE 55

lack of mannerisms. In use, such characteristics become a strong 'statement' in their own right: we quickly perceive that these are sound, strong, unassertive objects which use minimum material whilst also having elegant proportions and good finish. Their democratic inspiration does therefore communicate itself visually, but the aim of the designers and manufacturers was not just to create a 'look'. It was to achieve those in-depth characteristics.

Probably neither combination of designer and manufacturer would be insulted if one called these rather 'socialist' or even 'puritan' chairs. Their aim is to do something ordinary. The element that was 'not otherwise available' in these cases was their combination of mass accessibility with quality. So these are clearly also cases where the needs of the end-user provided the initial impetus and main motivation for creating the chair.

We are left with three motivations which concern the materials of which the chair is made. One case concerns choice of the actual raw material. Another is concerned with exploiting hitherto disregarded properties of a material that is familiar in this field. The third concerns introduction into the furniture production trade of new technologies, and hence advanced materials, with which it has never previously experimented. The novelties of form which result from these three kinds of innovation in materials may come to acquire psychological loadings amongst users, or satisfy their practical needs in new ways, but those are consequences, not the initial motivation. In these cases the impetus to develop a new chair is emerging from the concerns and interests of *production*, from a new opportunity which the manufacturer identifies in his own field, not from a new requirement expressed by or identified amongst users.

From our list of six questions, the dominant ones here are 'what is it made of?' and 'how is it made?' We have two left. Somewhere in the territory between the user's needs and the maker's concerns remain the questions: 'what shape is it?' and 'how do all the people concerned interact with the object, physically and mentally?'

21 SHAPE, AND HUMAN INTERACTION WITH THE OBJECT

You have been through enough discussion of these chairs by now to realise that *shape* derives on one hand from the properties and possibilities of a given *material* when treated in a certain way, and on the other, from the *practical and psychological functions* the object is intended to serve for the end-user. From whichever direction the designer starts, the task is one of synthesising these two sets of demands into a form that satisfactorily answers both of them simultaneously.

You will also have realised that every participant in the conception and life of this object in some way 'interacts with it, mentally and physically'. Into this category would come, for example, physical aspects of how the object is handled at every stage, including storage by the retailer and transportation to the place where the user will use it. But it also includes safety questions: is the object physically safe? Will it unambiguously explain to the user what it is and how it works? What precisely are the features that create certain psychological responses to the object, that cause it to have certain attractions or connotations to a given public, in a given culture?

It is not enough for the designer simply to recognise that the problem has these many dimensions. The designer has to know the facts that matter in each field (or where to find them). The designer has to know how those seemingly disparate fields interrelate in general, and be able to predict, if it is not already known from experience, how their particular options and components will impinge on each other. As I said before, it is a territory of infinite permutations and combinations, which is kept under control only by working carefully, logically and systematically, and by clearly distinguishing between what is known and what is speculative – to put it another way, between what might be seen as the 'science' of each of the sub-areas, and what is 'art' in the bringing together of them.

Even in this simple example of the chair, we have had examples of quite a number of these 'shape' and 'interaction' factors. Indeed those examples which can be culled from our selection of chairs will be the raw material for your own design exercise soon, in TMA 01. Our next stage, though, is to bring together all these different factors which we have discussed, into a manageable diagram, so that we can picture them as a whole.

22 BRINGING ALL THE FACTORS TOGETHER

All the elements of my diagram (Figure 56) will already be familiar to you from our discussion so far. At the top we have the two kinds of user function, physical and psychological: satisfying the needs of the body and the needs of the mind.

In the middle we have the questions of shape and materials. As we have seen, these are closely connected to each other, and choices in both areas are also directly influenced by the requirements of the physical and psychological functions.

At the bottom left, we have those detailed factors which must be taken into account in refining that basic shape into a fully effective and acceptable object. Here we are considering precisely how the object is to interact with people at every stage. Certain factors in this area, for example the precise placing of controls, or detailed decisions on colour, may sometimes have such primary importance to the practical or psychological functions that they are not 'detail', but are primary decisions better considered under those headings than here. In the case of a chair, these factors are relatively simple. Safety, on the other hand, appears simple but needs just as much careful consideration for a chair as for any other object.

At the bottom right, we have the other very detailed topic: production. How is the object actually 'produced', i.e. how is it going to be manufactured and brought into the presence of the user?

The commentaries in the intermediary boxes are just reminders (assuming you have read the Block!). They summarise what I have said about the logical connections between these aspects of the design task.

I suggest you read the diagram through from top to bottom a couple of times at least, in order to familiarise yourself with it fully.

If it reminds you of a cross between a systems diagram and a snakes-and-ladders board, that is very appropriate. If you follow the arrows around a bit, reading the contents of the boxes through which they pass, you should start to get a feel for that characteristic of designing which I mentioned in the first bit of Section 20. This is that it constantly involves links and relationships between objective factors and subjective ones, between the precisely quantifiable and that which can only be predicted somewhat speculatively. The arrows show, for example, that the designer has to face up to the *perceptual* aspects of shapes which it would be simpler to determine only by objective measurement. In that sense, this is a systems diagram: everything is connected directly or indirectly to everything else within the boundary, and the nature of the connections is indicated.

The snakes-and-ladders aspect will come to life when we use this diagram as a game-board for the TMA exercise in designing your own chair. All the instructions for that are in the Supplementary booklet.

Before you launch into the fun, however, I want to explain how this diagram and its contents relate to the rest of the course.

The aim of this introductory Block has been to introduce you to the key topics of design in a free-ranging way that will not only inform you, and forewarn of topics later to come, but will help you to clear your mind of restrictive preconceptions. This is essential to profitting fully from the material to come. It is equally important, whether you already work in some area of design, or whether you consider yourself (probably too modestly) to have 'known nothing before starting the course.'

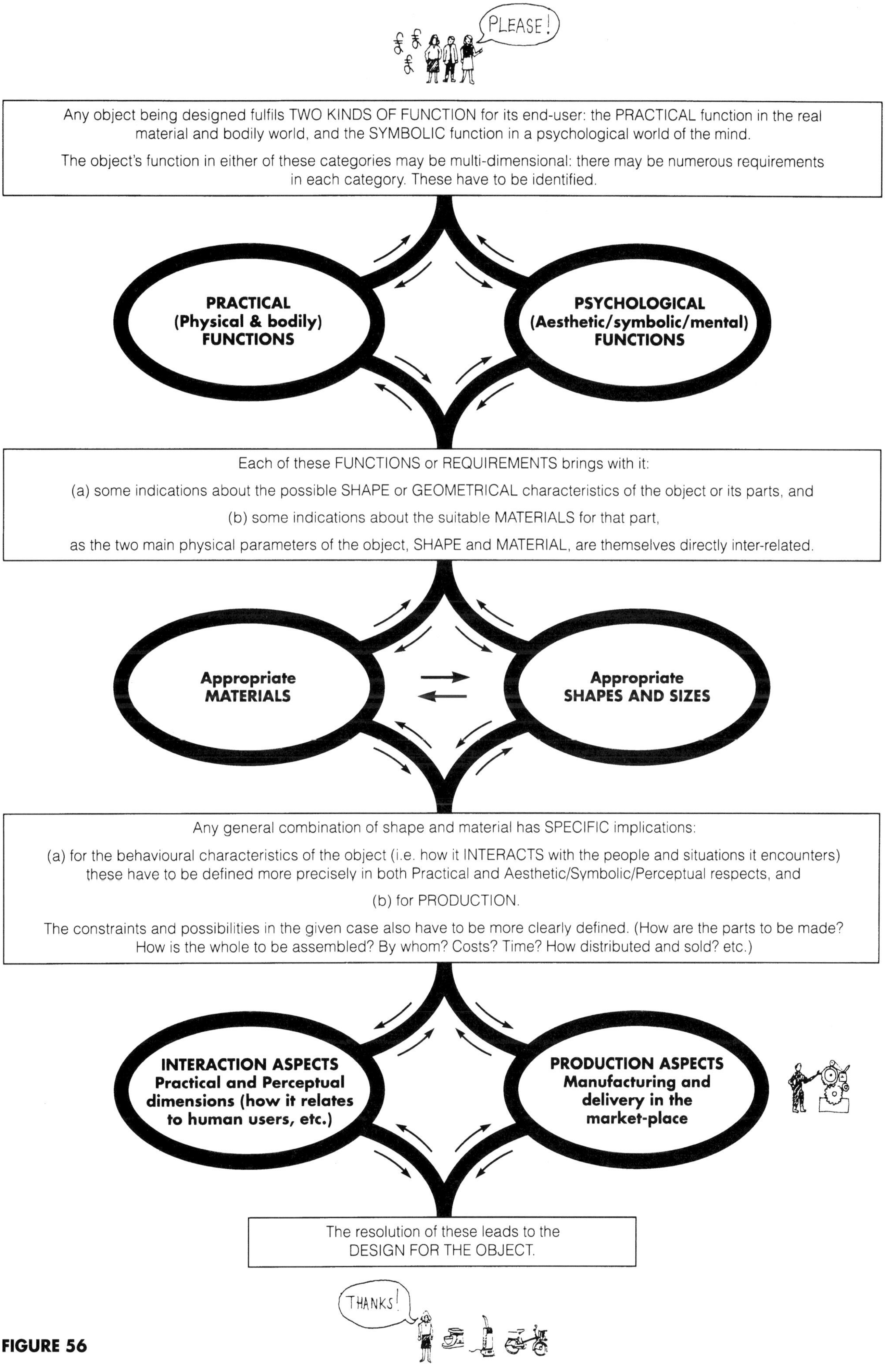
PLEASE!
Any object being designed fulfils TWO KINDS OF FUNCTION for its end-user: the PRACTICAL function in the real material and bodily world, and the SYMBOLIC function in a psychological world of the mind.
The object's function in either of these categories may be multi-dimensional: there may be numerous requirements in each category. These have to be identified.
PRACTICAL (Physical & bodily) FUNCTIONS
PSYCHOLOGICAL (Aesthetic/symbolic/mental) FUNCTIONS
Each of these FUNCTIONS or REQUIREMENTS brings with it:
(a) some indications about the possible SHAPE or GEOMETRICAL characteristics of the object or its parts, and
(b) some indications about the suitable MATERIALS for that part,
as the two main physical parameters of the object, SHAPE and MATERIAL, are themselves directly inter-related.
Appropriate MATERIALS
Appropriate SHAPES AND SIZES
Any general combination of shape and material has SPECIFIC implications:
(a) for the behavioural characteristics of the object (i.e. how it INTERACTS with the people and situations it encounters) these have to be defined more precisely in both Practical and Aesthetic/Symbolic/Perceptual respects, and
(b) for PRODUCTION.
The constraints and possibilities in the given case also have to be more clearly defined. (How are the parts to be made? How is the whole to be assembled? By whom? Costs? Time? How distributed and sold? etc.)
INTERACTION ASPECTS Practical and Perceptual dimensions (how it relates to human users, etc.)
PRODUCTION ASPECTS Manufacturing and delivery in the market-place
The resolution of these leads to the DESIGN FOR THE OBJECT.
THANKS!

FIGURE 56

Where this Block has introduced you to a topic that is treated in detail later, I have indicated that. Those links at the level of *detail* are quite straightforward. The slightly more complicated question is how our discussion here in Block 1 relates to the *structure* of the rest of the course. As this diagram represents a summary of Block 1, it is a question of how the contents of that diagram relate to the rest of the course.

I mentioned back in Section 20 that design necessarily always starts with a first stage of general exploration of the problem in all its inter-related aspects. This is partly a process of self-education for the designer, if he or she has not tackled a problem of this kind – say a chair – before. Or it may serve as a checklist and reminder of how this chair design problem differs from apparently similar jobs. In this stage, for example, the designer might be exploring the basic implications of the fact that the manufacturer who previously wanted a solid and hygenic hospital chair now wants a trendy chair for youth clubs. Or of a new client's desire to achieve a lower price than another designer could achieve for such a chair, and so on.

Whatever the degree to which the problem is a new one for the designer or design team concerned, this first period of exploring the territory will produce a 'concept'. This is a set of sketches, drawings or perhaps a model, defining the key characteristics of this object, its important features, and in particular defining and describing 'what's new about it'.

From this 'concept', the designer or team will move forward in conjunction with specialist colleagues or consultants, as I mentioned before, to work out the design in greater detail. Early consideration must be given to the precise form and geometrical arrangement of the parts, be it a piece of domestic equipment, an industrial machine, or a building.

With the results and conclusions of that stage, the next question to be examined is obviously the nitty gritty of what materials to use, how to process them into the components required, and how to assemble them all to form the 'object'.

You can see the simple logic of this, and imagine the sequence of stages taking place in a given company, even if you have no personal experience of it yourself. After the stage of 'thinking in all directions', which is the designer's initial contribution, other considerations of a managerial kind plainly come into play in order that work can flow sensibly through a company's various departments, or back and forth to its various consultants and subcontractors. To ignore such managerial or financial disciplines is plainly to threaten economic success, and hence the chances for producing this object at all.

Before all that even starts (as we have seen in discussing participants and their roles), there is information to be fed into the designer's thinking from market research, and feedback from criticisms of previous comparable products.

So here we have in a nutshell the structure of the course. It follows precisely this logical sequence through which virtually any product emerging in a marketplace today will have passed. To maximise the material available to you, and the number of points we can illustrate, each stage of this process is discussed through the medium of a particular type of product. Each has been chosen because it offers a particularly rich demonstration of the stage concerned. Here you see how the course develops in this order, with the type of product used as the main example in brackets:

Block 2: Market research and establishing the basic product specification (household consumer goods);

Block 3: Conceptual design and invention (bicycles);

Block 4: Geometry and layout (houses);

Block 5: Detail design and manufacturing (cars).

As I said, our last discussions here in Block 1, and the first TMA, give you not only an overview of the whole field of design, but a foretaste of the very first exploratory stages of that key process of 'conceptual design' to which we return in Block 3.

Turn now to the Chair Design Activity Supplement, and enter into the process of designing a chair in person. You will be referring back to this Block text several times as that process proceeds.

ANSWERS TO SELF-ASSESSMENT QUESTIONS

SAQ 1

(a) From the point of view of 'market pull', the form of a product is influenced by the needs and desires of potential customers/users. 'Needs' are those factors we term 'practical functions' (practical tasks which the customers or users seek to have fulfilled); 'desires' are those we term 'psychological functions' (symbolic tasks or aesthetic criteria which customers or users seek to have satisfied).

(b) From the point of view of 'production (or technology) push', the form of a product is influenced by questions relating to raw materials, production technologies, social, financial or legal pressures which cause the producer to introduce a particular new product into the marketplace.

SAQ 2

1 Scientific knowledge, describing how the world is, physically, socially, economically, etc. (e.g.: the conventional sciences like physics, chemistry, etc.; mathematics; mechanics; science of properties and behaviour of materials; geometry; statistics; economics).

2 Philosophical knowledge, describing what is important to people (e.g.: religious and other philosophy; politics; psychological and symbolic preferences; aesthetics).

3 Technological knowledge, describing how to handle the material stuff of the physical world to serve human purposes (all ways of making and shaping materials through applications of energy to them).

SAQ 3

My diagram of the relationships in the case of the birthday cake. Here, of course, no money changed hands, except when the mother, as maker, bought raw materials. As in the text diagrams, I have excluded all purchase of raw materials as this lies outside the boundary of the system that interests us here.

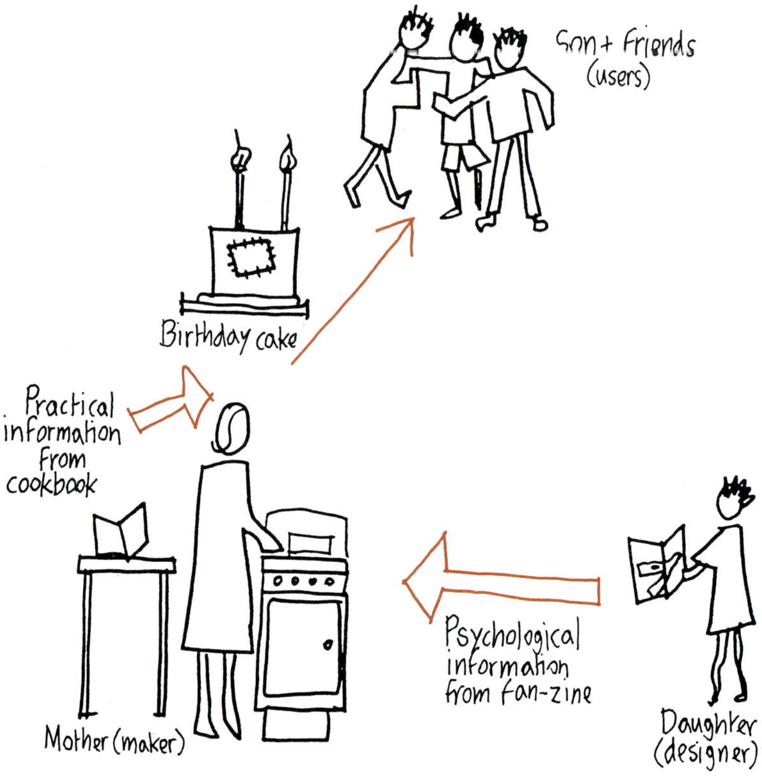

SAQ 4

My diagram for the case of the cordless vacuum cleaner, which is an example of the set of relationships represented in Figure 9(E).

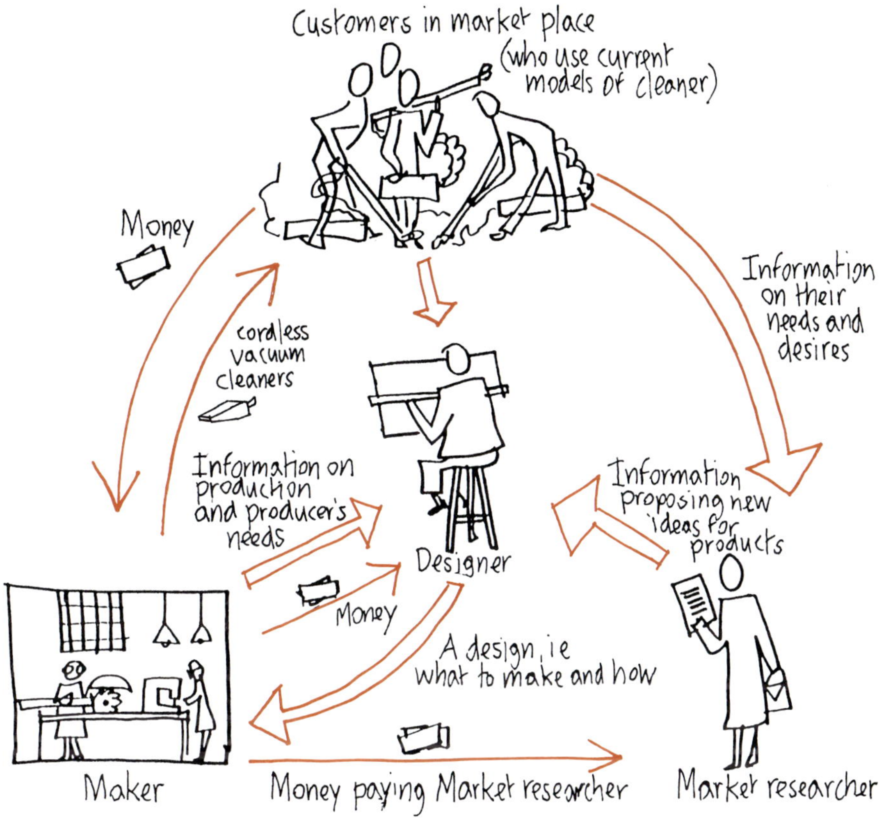

SAQ 5

The three main types of 'model' a designer uses to communicate are:

1. a three-dimensional model, in the conventional sense of the word 'model', that is a miniature replica of the object itself, or a full-scale mock-up;
2. a two-dimensional representation of the object as some kind of drawing, which might be a realistic picture (as for 'presentation', for example to a client), or a technical abstraction (as 'working' or 'engineering' drawings to instruct a manufacturer how to make it);
3. an entirely abstract representation in some form of numerical or other code, likewise for instructing the maker how to produce it (examples are knitting patterns; punched cards for controlling patterns in traditional looms; punched-tape or computer-programs controlling lathes or other industrial production equipment).

SAQ 6

1. Computer technology, in the form of computer-aided design and draughting systems, can produce drawings automatically.
2. Computer aided design systems can transmit coded instructions direct from the designer to computer-controlled manufacturing equipment, to assembly robots etc., which make the item to the designer's instructions without further human intervention.
3. Computer technology also makes possible life-like representations which can be animated to produce television-like 'simulations' of an object's operation, its movement, how it is assembled, etc.

SAQ 7

Despite selling well an object may be 'bad' either because it does not work, or not for long; or because it is dangerous.

SAQ 8

The most common way in which an objectively 'good' design proves unsuccessful from the maker's point of view is that it does not make enough profit; or more specifically: that it sells well, but cannot command a price high enough to yield the makers a profit.

SAQ 9

The issue is the morality of using the world's resources to produce 'useless' items to fulfil artificially created 'wants' in affluent countries, when much of the world's population in the other countries lacks for essential living equipment of the technically most primitive kind.

SAQ 10

The key question raised is: should objects be made just strong or powerful enough for their stated 'purpose', or should an object have reserves of capacity, strength, etc., over and above that necessary for its immediately and explicitly 'intended' purposes?

SAQ 11

The two main positive aspects of some built-in redundancy mentioned in the Block are:

1 the safety factor said to be created by having 'power in hand', or unused reserves of loadbearing, etc.;

2 the flexibility of being potentially usable for other purposes as well as that 'intended'.

The two main negative aspects mentioned are:

1 the hazard of putting significant reserves of power in the hands of untrained persons for use in unpredictable, uncontrollable and unsupervised conditions;

2 the legal difficulty which can arise in case of equipment failure, accident, etc., over what constitutes 'reasonable use' of the item, or its 'fitness for purpose'.

CHECKLIST OF OBJECTIVES

Having completed your study of this Block you should now be able to do the following:

PART 1

1 Describe how a new object, or another intervention in the material world of objects, may relate to non-material forms of intervention in human situations, and discuss the designer's role in researching the right overall mix to solve a given client's problem.

2 Discuss the concepts of 'incremental' and 'radical' change in relation to artefacts, and identify examples.

3 Discuss the concept of a 'type' in relation to objects, and how types may emerge.

4 Discuss the concepts of 'practical function' and 'psychological function' in relation to objects.

5 Give examples of some of the social uses to which symbolic and aesthetic aspects of artefacts are put.

6 Identify the different broad categories of impulse that can lead a producer to create a new product.

7 Explain how the forces of 'market pull' and 'production (or technology) push' affect product design.

8 Explain the difference between the 'information' which defines a problem, and the 'knowledge' and 'skills' which a designer brings to the solving of it.

9 Give examples of relevant general categories of such information, knowledge and skill.

10 Distinguish between those parts of a design problem in which a designer draws respectively upon scientific, philosophical and technological areas of knowledge.

11 Explain the different ways in which a designer's own value system may influence his work.

12 Explain the scope and limitations of formalised design 'methods' or 'routines' in relation to the overall range of problems arising within design as a whole.

13 Distinguish the various people and groups who participate in a modern mass-production operation when a new product is generated, and explain how this has evolved from the simpler relationships characterising craft technologies.

14 Distinguish the concepts of 'client' and 'customer' in relation to different modes of production, and identify their implications for the designer's task.

15 Analyse different case-histories in terms of these relationships and identify transactions taking place between their participants.

16 Describe the main types of three-dimensional, two-dimensional and abstract models used by designers to communicate their proposals to others, identifying the main characteristics and typical purposes of each; be able to recognise them on sight.

17 Suggest some of the kinds of change in types of model and modes of communication in design brought about by computer technologies.

18 Appreciate the role and significance of the tasks involved in 'design for manufacture' or 'design for assembly'.

19 State the three key questions to be asked before a drawing is prepared or other form of communication is undertaken.

20 Explain why the terms 'good' and 'bad' can generally only be relative in the evaluation of any particular design.

21 Describe the difference between 'objective' and 'subjective' criteria in evaluating the merits and demerits of a design, and indicate the types of characteristic to which each of these two kinds of criterion is appropriate.

22 Discuss the conflicts that potentially arise in reconciling 'good business' and 'good design'.

23 Define the concepts of 'needs' and 'wants' in relation to design, and explain how these relate to Victor Papanek's distinction between 'the real problem' in a society and 'the designer's share' of it.

24 Give a simple explanation of the concept of 'appropriate technology'.

25 Explain the concepts of a tailor-made 'fit' between an object and its 'intended task', and the alternative of 'functional redundancy'.

26 Identify some of the positive and negative implications of designing objects which offer some 'functional redundancy'.

PART 2

27 Attempt a simple analysis of familiar objects in terms of their fulfilment of practical functions, and their social or symbolic specificity to a given culture, using chairs or other examples.

28 Give examples of how the shapes of an object's components, the way the components are assembled, the selection of materials or of the colours in which they are finished, can be a means of expressing cultural or philosophical values, of creating symbols, or of otherwise enabling the object to fulfil 'psychological functions' in its particular culture.

29 Describe some of the kinds of event or factor which may typically generate new designs, with reference to examples in the history of chair design during and after the Second World War.

30 Describe some of the ways in which the same idea may come to be circulating in different parts of the world at about the same time.

31 Explain the basic principle of 'reinforcement' of one material by another.

32 Outline some of the methods by which timbers are processed in application to furniture, and the characteristics each offers.

33 Give simple explanations of the characteristics of shape or geometry that typically make a form stiff or springy.

34 Make simple classifications of materials, into those which are 'three-dimensional' (i.e. have bulk), 'two-dimensional' (i.e. are of sheet form), and 'one-dimensional' (i.e. consist of long lengths of relatively very small cross-section); indicate in general terms the main means of shaping materials in each of these three categories.

35 Explain the general concept of 'semantics' with reference to design, and distinguish the two semantic techniques of 'representation', i.e. communicating a meaning by some literal quotation, replication or naming of another object, and 'evocation', i.e. communicating a meaning by indirect reference such as sensual evocation or subtle intellectual association.

PART 3

36 Explain the concept of 'standards' and describe some of their uses.

37 Describe the most common kinds of distribution system that may operate between producers of objects and their customers or users, and the different roles they may play beyond the purely mechanical one of getting goods from producer to user.

38 Give typical examples of the kinds of factor that may form the 'initial impetus or main motivation' for a design process to be set in motion.

39 List the six main categories of factor to be taken into account as the process of design proceeds, and indicate some of the questions they typically raise for designers to answer.

40 Describe in simple terms some of the logical linkage which a designer is following between these six main categories of factor, and explain why the path followed, and number of iterations will be different in almost every case.

VIDEO 1 SECTION 1 'PEOPLE AND MECHANISMS'

41 Define the concepts of 'independent mechanism' and 'human contact surface' as developed in the video.

42 Analyse a familiar object in terms of these two concepts.

43 Classify objects into the broad categories used in the video, according to their scalar relationship to the human being who interacts with them.

44 Explain the significance of electronic miniaturisation of mechanisms for the designer's approach to 'practical' and 'psychological' functions of an object.

ACKNOWLEDGEMENTS

Grateful acknowledgement is made to the following sources for permission to reproduce material in this Block:

Text

p.45: Papanek, V. (1972) *Design For The Real World*, Pantheon Books.

Figures

Figure 1: N.A.S.A.; Figures 2(a) and 2(b): Hulton Deutsch Collection; Figure 2(c): Institute of Agricultural History and Museum of English Rural Life, University of Reading; Figure 3(a): National Motor Museum, Beaulieu; Figure 3(c): Quadrant / Autocar / AutEx; Figure 4(a): Banham, R. (1960) *Theory and Design in the First Machine Age*, Butterworth Heinemann; Figure 4(b): Meller, J. (1970) *The Buckminster Fuller Reader*, Jonathan Cape. Also reproduced by permission of Bantam Doubleday Dell; Figures 5, 14, 15 and 16: Papanek, V. (1972) *Design for the Real World*, Pantheon Books; Figure 7: Courtesy of David Brown Radicon Ltd; Figure 8: Jenkins, J.G. (1981) T*he English Farm Wagon: Origins and Structure*, David and Charles; Figures 10(a), 10(b), 10(c), 10(d), 10(e) and 10(f): Courtesy of Olivetti Office Ltd; Figure 11(b): Courtesy of Univac Built-In Vacuum Systems Ltd; Figures 19, 40 and 48: Logie, G. (1947) *Furniture From Machines*, HarperCollins; Figure 20: From the catalogue of William Whiteley Ltd, in Schaeffer, H. (1970) *Nineteenth Century Modern: the functional tradition in Victorian design*, Henry Holt and Co; Figure 22: Courtesy of Nada-Chair; Figures 23 and 28: Ishimoto, Y., Gropius, W. and Tange, K. (1960) *Katsura: Tradition and Creation in Japanese Architecture*, Yale University Press. Reproduced in Alex, W. (1963) *Japanese Architecture*, Prentice-Hall International; Figure 24: Crafts Council; Figure 25: Seike, K. (1977) *The Art of Japanese Joinery*, Weatherhill / Tankosha; Figure 26: Thomas Sheraton's *Cabinet Dictionary, Volume 1*, reprinted by Praeger Publishers, 1970; Figure 27: Weidenfeld and Nicolson Ltd., Kerry Dundas; Figure 29: Gloag, J. (1964) *The Englishman's Chair*, Allen and Unwin; Figure 30: Popperfoto; Figures 31 and 32: Geffrye Museum; Figures 33, 34, 35, 37, 38, 39, 45 and 49: By courtesy of the Board of Trustees of the Victoria and Albert Museum; Figure 36: Mr Abram Games and G and B Arts Ltd; Figure 41: Gebruder Thonet Arktiengesellschaft; Figures 43(b) and 43(c): Artek; Figures 44(a), 44(b) and 44(c): Kandya Ltd., photo Whitecross Studios (John Cook), London; Figure 46: E. Kold Christensen, photo K. Helmer Petersen, Copenhagen; Figure 50: Peter Fiell; Figure 51: The Trustees of the Edward James Foundation; Figure 52: Atelier Hans Hollein.

Figure 3(b) was provided by Mike Levers of the Open University Photographic Studio, and Figures 2(d), 2(e), 11(a), 11(c), 12, 13, 17, 18, 21 and 42 are courtesy of the Block author, Catherine Cooke.

T204 DESIGN: PRINCIPLES AND PRACTICE

COURSE TEAM

Academics	Godfrey Boyle Catherine Cooke Nigel Cross James Forster Georgy Leslie Ed Rhodes Joe Rooney Robin Roy George Rzevski Philip Steadman David Walker
External Assessor	Michael Tovey (Coventry University)
Consultants	Michael Baker Stephen Brown Barry Dagger Susan Hart Jim Platts
BBC Producers	Cameron Balbirnie Ian Spratley Bill Young
Course Manager	Ernie Taylor
Editors	Keith Cavanagh Garry Hammond Rodney Wilson
Graphic Designer	Rob Williams
Graphic Artist	Keith Howard
Media Librarian	Caryl Hunter-Brown
Secretaries	Carole Marshall Margaret Barnes Jennie Conlon Pat Dendy